ALSO BY MATTHEW GAVIN FRANK

NONFICTION

*Flight of the Diamond Smugglers: A Tale of Pigeons, Obsession, and Greed Along Coastal South Africa*

*The Mad Feast: An Ecstatic Tour Through America's Food*

*Preparing the Ghost: An Essay Concerning the Giant Squid and Its First Photographer*

*Pot Farm*

*Barolo*

POETRY

*The Morrow Plots*

*Warranty in Zulu*

*Sagittarius Agitprop*

# SUBMERSED

# SUBMERSED

Wonder, Obsession, and Murder in the
World of Amateur Submarines

## MATTHEW GAVIN FRANK

PANTHEON BOOKS
*New York*

FIRST HARDCOVER EDITION PUBLISHED BY PANTHEON BOOKS 2025

Copyright © 2025 by Matthew Gavin Frank

Penguin Random House values and supports copyright. Copyright fuels creativity, encourages diverse voices, promotes free speech, and creates a vibrant culture. Thank you for buying an authorized edition of this book and for complying with copyright laws by not reproducing, scanning, or distributing any part of it in any form without permission. You are supporting writers and allowing Penguin Random House to continue to publish books for every reader. Please note that no part of this book may be used or reproduced in any manner for the purpose of training artificial intelligence technologies or systems.

Published by Pantheon Books, a division of Penguin Random House LLC, 1745 Broadway, New York, NY 10019.

Brief portions of this work were previously published, in different form, in *Harper's Magazine*.

Pantheon Books and the colophon are registered trademarks of Penguin Random House LLC.

Library of Congress Cataloging-in-Publication Data
Names: Frank, Matthew Gavin, author.
Title: Submersed : wonder, obsession, and murder in the world of amateur submarines / by Matthew Gavin Frank.
Description: First edition. | New York : Pantheon Books, 2025 | Includes bibliographical references.
Identifiers: LCCN 2024026052 (print) | LCCN 2024026053 (ebook) | ISBN 9780593700952 (hardcover) | ISBN 9780593700969 (ebook)
Subjects: LCSH: Wall, Kim, 1987–2017. | Madsen, Peter, 1971– | Murder—Denmark—Case studies. | Submersibles—Social aspects—History. | Submarines (Ships)—Psychological aspects.
Classification: LCC HV6535.D4 F73 2025 (print) | LCC HV6535.D4 (ebook) | DDC 364.15209489—dc23/eng/20250213
LC record available at https://lccn.loc.gov/2024026052
LC ebook record available at https://lccn.loc.gov/2024026053

penguinrandomhouse.com | pantheonbooks.com

Printed in the United States of America
2 4 6 8 9 7 5 3 1

The authorized representative in the EU for product safety and compliance is Penguin Random House Ireland, Morrison Chambers, 32 Nassau Street, Dublin D02 YH68, Ireland, https://eu-contact.penguin.ie.

*For Louisa*

NOTE: The names of persons mentioned herein who have not previously been a part of public records or discussion have either been changed where noted or redacted.

# SUBMERSED

# PROLOGUE

For as long as I can remember, I've been afraid of the ocean. I've had drowning nightmares since age four—nightmares that have persisted into adulthood with little variation: I'm standing waist-deep in the ocean. Onshore, my mother stretches out on an orange chaise longue beneath an orange beach umbrella, sipping chocolate milk from an orange thermos. In the sky, smoke snakes from an orange propeller Cessna plane. It's trailing one of those banners behind it, advertising a sale on oranges at some local grocery. When the undertow sweeps my legs and sucks me farther out, I try shouting but water rushes into my mouth. I sink, surface, sink, surface. The world is muddied, the plane has crashed behind some luxury hotel, and my mother is standing, waving her arms over her head to signal to me, or to signal to someone to help me. The banner flutters in the air like an eel, shrouding the sun, dropping to beach grass. The luxury hotel bursts into flames. My mother clutches her thermos, and the chocolate milk is flying from it. I go under and I don't surface. My body feels as if it's going to explode. An eel slithers by, slows, and stares at me unblinking as I wane. The world goes orange. I wake gasping.

I've spent much of my waking life fixated on the ocean from afar. I love watching it like I love watching a campfire—attracted, but stopping short of leaping in. My mother nearly drowned as a child, swimming at Rockaway Beach. Her father—a used car salesman who was not a strong swimmer—had to dive in, save her life, and one week later he died in a car wreck. She never swam in the ocean again, and still can't bear to look at it without getting anxious. She told me that she believed the ocean had something to do with her father's death. "It's silly," she said, "but if he had just let me drown, he would have probably lived a longer life." I was a boy when she told me this, and she was comforting me after I had woken up, screaming, from my own ocean nightmare. I had sweat through my *Star Wars* sheets. She sat on the edge of my bed, combing her fingers through my hair. She didn't just fear the ocean. She *distrusted* it, felt it was conspiring to do harm to us. She sat with me until dawn, until the woodpecker we always heard but never saw began feeding from the siding outside my bedroom window, like a metronome. To me, my mom's dad—my grandfather, I guess—was another legend I'd never know. A cautionary tale. A warning. "Don't be afraid," my mom said, but I didn't believe she meant it. She was reciting lines. Her heart wasn't in it. Somehow—whether narratively or via the hiccups of heritability, or via her fingers through my hair—she has sewn that fear, and distrust, into me.

The manifestations of my anxiety are often stubbornly similar to my mother's. But I've tried to find ways to engage my fixations—especially those that have roots in a phobia. I come from a long line of people who suffer from overwhelming OCD. I've channeled mine into becoming obsessed with obsessives—burrowing into the machinations and secret desires of one niche community or another in order to find out something about us, our evolving narratives about the human condition, and the ways in which those narratives sometimes gel with and sometimes contradict our actions; to uncover or restore some bewildering intricacy to that which drives and defines us as a species.

≈

In August 2017, the Swedish journalist Kim Wall was brutally assaulted and murdered by the amateur submersible builder Peter Madsen while on board his sub *UC3 Nautilus* off the coast of Copenhagen, Denmark. I did not want to write about the murder of Kim Wall. I wanted to engage the eccentric microcommunity of DIY submersible enthusiasts, and to scratch at their obsessions and their actions for some larger—if elusive or illusory—meaning; some sly microcosmic comment on the human condition and on human longing. I wanted to write about the flights of fancy and motivations of the PSUBS (or "personal submersibles") collective—a group of self-proclaimed free spirits who identify first and foremost as "amateur submarine builders and underwater explorers." I wanted to write about coral beds, and about bioluminescence.

But in my research I kept bumping up against violence, misogyny, the murder of Kim Wall. I couldn't help but confront and interrogate the inflection points at which a sense of wonder sours into something more malign. When and why and how does the compulsion to sink to depth uncannily begin to dovetail with darker, more threatening traits? What is it with my own fraught compulsion to sink to these depths, the ethics of inquiring into and interacting with—in book form—real-life atrocity?

What is it to engage with this thread of "true crime" from this vantage? To realize that an engagement is ethically unsound, but to be unable to stop oneself from pursuing it, is to acknowledge that (to simplify) I'm not a very good person. And even if one attempts to desensationalize the grisly (which, of course, is not innately sensational but becomes so only if passed through the lurid portal of a particular breed of reportage), is one's initial draw to the subject matter rooted in part in the sensational?

I passed many a midnight combing through court documents pertaining to the Wall/Madsen case, translating testimony and transcription, rereading the articles—from Danish, Swedish, Norwegian, British, and American newspapers and magazines—noting how the angles and the tones shifted from the early reportage to the aftermath, fresh, excited hot takes arising with each new bit of evidence revealed, then curdling when the implications of such revelations settled in a few days on. I read articles and watched early documentaries on Peter Madsen that predated the murder. I learned about his upbringing, his parents, his friends, his evolving hobbies. I read and watched interviews of many of the people who appear in this book. I watched them move, watched their faces and hands, recorded the moments when their voices rose or fell, sped up, slowed down, broke, and went silent. I recorded when they seemed to be excited, surprised, confused, exasperated, and sad; baffled by their own actions or by the things they found themselves loving or fearing.

I read everything written by Kim Wall that I could find, until I became obsessed with some of the things that obsessed her as a writer. I read everything written *about* Kim Wall that I could find—not only by journalists covering the story of her disappearance but by her friends and family and colleagues. I scrutinized photos and videos of Kim; of Madsen; of Kim and Madsen together on that night. I fixated on the subjects of the photos and videos, and then I fixated on their corners, edges. I wanted to see what was lurking above or below or behind the people in these images (and, in a way, that which was lurking above or below or behind the reported facts and testimonies). I tried to pinpoint the spaces in which they were taken or filmed, researched what lay outside the frame, the larger, looming physical context that shaped these moments—the smiling or scowling or worried people living them. And then, when this wasn't enough, I traveled to the places where the pictures were taken, the videos were filmed; the sites and towns and buildings and workshops. I retraced the same paths once walked by the people about whom I was writing.

Desperate to appraise the intensity of Madsen's interests, I started seeking the voices of other DIY submersible enthusiasts, sometimes traveling great distances to sit with them in their living rooms. I was hoping to make sense of this odd, niche obsession and its relationship with tragedy, and the ways in which a sense of wonder—wild-eyed and often joyous—mitigates the nagging awareness of the other shoe that often seems to drop into or onto the lives of these enthusiasts, and those who choose to get close to them, even if just to chase an interesting story. In these interviews, in desperately attempting to uncover the why behind the obsession, I was hoping to illuminate—however deficiently—the why behind the murder; to break down the cocktail and isolate its individual ingredients. I found myself wondering how I was implicated in all of this—how my own interest in the DIY sub community may have been gathering its own looming tragedy. How my urge to speak to as many of these folks as I could usurped my urge to avoid potential harm, blinded me to it.

In engaging such a story, there are inevitably moments wherein the evidence and the recorded facts fall short, and frame a kind of missing, ultimately unknowable center: What was Madsen really thinking and feeling in the moments leading up to the murder; what was Kim thinking and feeling as she walked to his laboratory that night to pursue the article she wished to write about him? In grappling toward that unknowable center—itself an admittedly dubious act—I clutched at the mosaic of facts that I was able to compile, and arranged and rearranged them around these unknowables to see how each could lend resonance to the other.

When writing of Madsen's feelings, I based his "inner life" on a combination of testimony he himself later gave, mapped over and onto that which his friends and interns had to say about his demeanor and text messages he sent, as well as interviews he gave just hours after the murder took place and the journalistic responses to and framing of his words and comportment. When writing of Kim's loved ones during this time, I culled from police reports in which they were interviewed, photographs, text messages (that were made available to the public), newspaper articles detailing their conversations with Danish naval officials, and narratives that they themselves wrote.

In every act of reportage or essaying, of course, is an act of framing; and framing is a conversational act, an inescapable interaction between author and subject. Bound to a body and the seething chemicals therein, every author is a medium; a filter rife with frailty, however responsible and ethical they may fancy themselves to be. Even a sober tone of journalistic dispassion requires an act of conjuring, performance. When a fact passes from the world through an author, to a reader it can sometimes emerge a little kinked, because it took some of the author with it on its passage. And when all we have is language—sentences—by which to express these stories, we have only a faulty tool. "Words are our weakest hold on the world," says the poet Alberto Ríos.

Speaking of research, the often brilliant and sometimes exasperating sixteenth-century French essayist Michel de Montaigne said, "It is good to rub and polish our brain against that of others." The contemporary New York City essayist Leslie Jamison believes that research can "sensitize [us] to points of connection." In obsessively gathering and subsequently interacting with every fact, testimony, interview, picture, video, and firsthand observation or act of witness, I tried, via this brain-rubbing, to make the frame as structurally sound as possible. And when I didn't, or couldn't, know something, I culled from crumbs that I did know, and arranged them around that which I didn't, in an attempt to exhume these instances of sensitization, drawing, however cryptic the metaphor, a sort of outline around that which remained fugitive; laboring to uncover the right blend of chalk necessary to evoke the body.

I worry about the fabric of the genre within which I find myself drifting, its mercurial nature, and what it needs to tell itself in order to sustain itself. Even if I scrupulously try to avoid the ugly luridness of the coverage of the Wall/Madsen case, even if I try to highlight new truths about it by focusing on the ephemera dancing around it, I am still likely doing harm to the victims—to Wall's family.

I realize that, even before the details of Wall's murder began attending, like metal flakes to a magnet, my investigations into and obsessions with the personal submersible community, my particular cage of skin lent me the privilege that mitigated the dangers Wall had to consider, negotiate, face. I had the privilege to be less complicatedly excited about diving to depth with one of these PSUBS guys, who is, in part, I suppose, a guy like Peter Madsen, or a guy akin to who a journalist may have thought Peter Madsen was—a hobbyist eccentric—before the murder occurred.

I can only hope—however inadequately, and perhaps in moments even infuriatingly—to draw inspiration from Wall's own work as I negotiate the threads of my own. "As the empathetic reporter she is," the journalist Sruthi Gottipati says of Wall, "she writes with nuance and gives her characters, however downtrodden or peculiar or disenfranchised, agency. I wish we lived in a world that does the same for her." I share this wish, and also recognize the shortcomings of my attempts to empathize with experiences that are not and cannot be my own. The fact may be that my unbridled chasing of my obsessions into the personal submersible world, and the seemingly ancillary threads attending that chase, is itself an ethically unsound act. I am sinking through, at great speed, alternately attracted to and repelled by the subjects with which I collide, but, regardless, I can't stop moving forward, or down.

# PART ONE

# 1

IF DENMARK WERE a human body, the island of Refshaleøen would be the appendix. It dangles into the Øresund Strait as if some vestigial appendage, subordinate to the larger island of Amager—to which it is presently annexed. On maps, the island appears a little obscene and expendable. Its streets front vast and empty warehouses, factories. The cold sea air has crept into these buildings' bricks and blocks and beams, compelled them to buckle and bloat. They resemble the petrified remains of giant sea creatures—their beached mythological rib cages frozen in mid-heave.

Many in Copenhagen still refer to the island as "a place to be transformed" from its long-ingrained status as crumbling defunct shipyard to (according to the island's town council) "a vibrant new destination." To that end, on the cold grounds of windowless slate-gray buildings that once held ship guts, fetid ropes, and rusting propellers, artists now paint and sculpt. Inspired perhaps by the warehouses' gaping spaces, their work is oftentimes larger than life and can't fit through the doors. This work is either imprisoned in the space in which it was created or must be dismantled into its parts. Eccentric engineers joint disparate found components—outriggers, hulls, shards of glass, conveyor belt ramps, oil cans, toilet seats, accordion bellows, and mannequin heads—into unholy machines that clank and whir, beautifully without purpose. Here, amid the decaying factories, there are now festivals—the Copenhell heavy metal extravaganza with its Hades stage; the Copenhagen Distortion Celebration of Nightlife with its "Anything Goes" credo sponsored by Adidas and Red Bull; and the Scandinavia Reggae Festival, "presenting artists from the Scandinavian reggae scene" such as Svenska Akademien, Fastpoholmen, and Spöket.

Many locals who lived through the oil crisis of 1973, which resulted in a global shipbuilding catastrophe, struggle to absorb this new narrative, this quick and drastic recasting of Refshaleøen's identity. They remember too well the oil embargo that spawned a price increase of nearly 400 percent in Europe. They remember how the shipyards in China, Japan, and South Korea capitalized on the severity with which the crisis hit Europe via a multifaceted plan involving wage cuts, state subsidies, streamlined production plants, and a keen price dumping agenda. Almost overnight, the European shipbuilding market share dropped from 41 percent to 18 percent, as the Asian market share jumped from 46 percent to 70 percent. They remember how, when Refshaleøen's Burmeister & Wain shipyard went bankrupt and closed its doors, that meant over eight thousand layoffs.

Those who once worked in the shipyard's many divisions—pipe production, design, engineering, carpentry, scaffolding, electronics, sandblasting, iron, and steel—were, according to studies, simply "absorbed," or "dissolved," meaning they were never traced, which is not the same thing as disappearing. All those spaces, which once held the chorus and noise of eight thousand people and their tools, their banging, their schematics, their hulls, and their sparks, every conversation shared, every confession of joy and tragedy, plans made and failed plans, suddenly were silent, abandoned, a few broken beams left behind, a few handfuls of rusty nails. The warehouses became lost colonies, and the wind whistled through them.

≈

A tree falls in the woods. A lightbulb inexplicably bursts in an empty warehouse. Ants the size of pinkies rove among the glass. Five kilometers away, at the central station, freight and passenger trains whine and rumble, cars couple and detach; wooden sleepers rattle, concrete sleepers vibrate, loads of coal topple and loads of sugar beets roll. The shells of the cars are uniformly damp and gleaming, and will dry and dull only after leaving Copenhagen, chugging along the rails on their way to some other sector.

The winds off the North and Baltic Seas meet in the middle of the Øresund Strait, and, even in the summertime, crust Refshaleøen with sea spray and chill. In the winter, the place is ice-hardened and dark, squeaky with cold. The sun sets before 3:00 p.m., and the eastern currents, carried to the island's shores by the Siberian High, can cause one to gasp. The ants go deep underground. It's too cold for the local teenagers to spray-paint their names and the names of Norse gods on the sides of the warehouses with ungloved hands. In their mittens, they fumble with the cans. Their faces are masked in the fog of their own exhalations. Half of their paint streams freeze in the air before reaching the warehouse flanks. Some of the frozen red mist spatters their boot-tops. Still, shivering, they manage to write their names amid the other graffiti—the lightning bolts and boobs, the hearts and the swastikas.

During these winter months, many locals swear by phototherapy devices—artificial light visors, or golden-lensed sunglasses to lend their world the illusion of brightness—to combat seasonal affective disorder. All this cold goes golden. All this ice, richly piss-colored. Refshaleøen and its shipyards and shop fronts and residents and sea become uniformly jaundiced. Such measures, though, often fail, and insufficiently mitigate the symptoms associated with living here during the winter—feelings of despair, hopelessness, dread, and anxiety, low energy, insomnia, apathy, and an almost unbearable obsession with death. The city council urges the inhabitants to lock up their firearms and put the keys in difficult-to-reach places, so they may have to think twice and exert sufficient effort before turning said firearms on their loved ones, or on themselves.

～

The Øresund Strait, with its maximum depth of 131 feet, conceals the invasive and omnivorous pacu fish that proliferate in its waters, and that are "fully capable"—according to the University of Copenhagen's cautionary article "Danish Swimmers Escape Waters Fearing Killer Fish"—"of severing fishing lines and even fingers." On their bellies are serrated keels. The pacu's teeth, disarmingly, look exactly like those of a human. They often have a slight overbite. With these teeth, they have chewed off the faces of other fish, have eviscerated juvenile crocodiles, have left scars on human shoulders, necks, and rib cages, scars that are often mistaken for human bite marks; scars that some joke are the result of overzealous former lovers. Scars that, especially in the winter months, still redden and itch. Pacu corpses sometimes wash up onto the shores of Refshaleøen, and children dare each other to open the fishes' mouths and not scream when they behold those humanoid teeth.

The pacu is endemic to the waters of South America and shouldn't be here in Denmark. How it got here remains a mystery, or a miracle (though few here would call the pacu's presence miraculous). Was it deliberately loosed into these waters by some traveling prankster? Was it a hardy escapee from a local's exotic aquarium? Did it slip through some rip in the space-time continuum, a phantasmal hopper amid biomes? Is its presence an anomaly or abomination, or both? Either way, the local wildlife authorities advise those who unexpectedly catch a pacu to immediately "cut the head off the fish and dispose of [it] as garbage." An abomination is a miracle that bites.

These waters have seen their fair share of so-called abominations. In 1550, the fabled sea monk washed up onto the shore of Refshaleøen and was classified as a monster whose presence signified that the island was cursed by the gods. According to the marine ecologist and author William M. Johnson, stories of the sea monk "were perpetuated with viral-like efficiency," and the legendary beast soon inflamed the fears of the European mariners of the era, some of whom claimed to have spotted the creature from the decks of their ships. It was described as a ghoul with "a human head and face, resembling in appearance the men with shorn heads, whom we call monks because of their solitary life; but the appearance of its lower parts, bearing a coating of scales, barely indicated the torn and severed limbs and joints of the human body." Naturalists today remain at a loss as to this creature's true identity. Was it a giant squid? A seal? A walrus? The legendary basilisk that could kill a human with a sideways glance? A magical Jenny Haniver—a confounding man-made sculpture fashioned from the dried and conjoined carcasses of a skate, ray, guitarfish, and/or angel shark, sewn together and manipulated to approximate an evil creature of myth—a dragon or demon or devil? The Jenny Haniver was said to possess terrible magical powers. Who made it? What was to be the nature of the hex?

On the quay, preparing to defrost, are frozen spit, frozen tears, frozen blood. The waters remain somber, and one may worry about the things they continue to dislodge. In 1983: the ancient ax blades of a sunken Mesolithic settlement dating from 6400 BC. On February 29, 2016: a dead puffin slumped over the fluke of a dead fin whale. On Monday, August 21, 2017, three months after the Copenhagen Distortion Celebration ("Scandinavia's wildest street party"), two months after the Copenhell heavy metal festival, and four days before the Scandinavia Reggae Festival: a woman's headless, limbless body.

# 2

If we're to believe Dr. Ernest Campbell, surgeon and specialist in "diving medicine," people who are obsessed with sinking to depth in large bodies of water "have different chemistries and personalities" than those of us who are not similarly obsessed. "Because of the effects of various gases under pressure . . . [they] respond differently to abnormal physiological states and changes in their environment." Our liquid and solid worlds, our lives at depth and our surface lives, can become confused. Our pressures are in flux, too much gas or too little gas affecting the ways in which we see and feel our worlds.

What makes a diver more unstable: The descent or the return? The dream itself, or the cruel sharpness of the world after waking? No matter: either way, the water, at depth, can ruin a person for the surface. When diving to a depth of one hundred feet, for instance, the body endures four times the pressure it's used to, and our organs begin to writhe, our lungs wrung as if sponges. We struggle for air. Still, we can't rise too fast, or the pressure will paralyze us, swell our brains. See all that light up there, convulsing like an amoeba? Hear all that noise about to unmute itself, bully our ears? That's madness.

≈

Refshaleøen's shipping containers make a colorful labyrinth, each outcompeting the other in the race toward decay. Stacked, they form kaleidoscopic walls of turquoise, orange, yellow, and cream, their flanks festooned with sunbursts of rust. Immersed in their maze, crunching the gravel and broken glass, the smell of the ocean remains so strong it stings the nostrils, waters the eyes. The air here—and everything it touches—stinks of salt and motor oil and viscera. It reddens the throat. Beyond the containers, is that a forklift backing up or a seagull screaming? If one is to successfully emerge from these shipping container lanes, one must follow the wailing sound.

It's arresting, stepping out into the expansive old shipyard. The defunct smokestacks in the distance are taller and skinnier than the average smokestack, white with black tips, giant noirish cigarettes abandoned unsmoked. They hover like sentries over the post-industrial establishments, wreckage repurposed into hot spots made of glass and steel and thrumming with chill-out electronica. Here, diners can eat dishes called "impressions" while sitting on their hands and wearing 3-D glasses. There, hedonists can don sensory deprivation helmets while immersing themselves in private Jacuzzis called "pots." When the steam and the deprivation become too much, one can pop off the helmet with the aid of the "ejector button" and emerge into vaporous Refshaleøen, to trance music and a nice cool bottle of sparkling rosé. One can don army fatigues in the Paintball Arena, housed in a converted ship storage warehouse, wherein "adrenaline junkies" can "let their inner sniper loose" and opt to use only red paint, which looks "just like real blood!" Once, Europe's largest vessels were built in this place.

Many of these football-field-size warehouses have been converted into specialty schools, offering classes in gourmet cooking and surfing, adult education and vehicle repair, pole dancing and psychotherapy. Students whisk and wax, read and screw, spin and analyze beneath sixty-foot ceilings, swaying fluorescents, exposed metal beams, the permanently closed inner doors of the four-story garages. Their voices echo in all that space. They return to their harborside modular homes with shared bathrooms in a "cost-worthy co-living village," clean-lined boxes of steel and glass nestled together to form narrow alleyways from which one can see only the tops of the nearby masts rising over the upper apartments. The "village" publishes its own journal, espousing the ethics of "living small" and the "science of co-eating" in which residents aspire to commune with the practices of "the Roman Army." The waiting list for an apartment is yearslong, but applications are always open.

Adjacent to the dance studio, around the corner from the defunct railroad tracks, and down the wet gravel street from the Port 4130 Skatepark, is the oxidized warehouse that houses Copenhagen Suborbitals' amateur space program, where hard-hatted and safety-goggled crews—ranging from blacksmiths to rocket scientists—scurry along their assembly lines in reflective vests building not boats but spaceships using, they must admit, "low-tech production methods in the workshop, [instead of] expensive, exotic materials." The company pays their electric and heating and water bills, as well as for their tools, alloys, and rocket fuel, with crowdfunding donations.

The company was founded in 2008 by the fringe artist and amateur submarine builder Peter Madsen and the avant-garde rocket maker and "space architect" Kristian von Bengtson (whose official brief biography stipulates, "He is only satisfied if a challenge is close to impossible"). Madsen developed a passion for engineering while in secondary school after bonding with his chemistry and physics teacher, who was the first to introduce him to the properties of rocket fuel. As a boy, Madsen was taken with the fuel's volatile characteristics, and, with his teacher's guidance, he began experimenting with the propellant. Such experiments empowered him. He wanted to see how far he could push the fuel, maximizing its propulsive effectiveness without pushing its components (and the reaction between them) toward a disastrous instability. Increasingly, he felt that he could tame the stuff. His teacher praised him as "very lively" and "exceptionally curious," commending him for his daring and encouraging his innovation. Madsen had very few friends his own age.

On his walks to and from school, Madsen would often take detours so he could be close to the sea. He would daydream about exploring its depths, which struck him as nearly as exciting as those promised by outer space, but so much more accessible. He felt that the sea could offer him an escape from the social structures maintained by his schoolmates, many of whom ostracized him, according to a former peer, "as something of a nerd." Increasingly, Madsen felt that the sea might be a space wherein he could indulge his passions and his desire to be a loner, without the scrutiny or contempt of his cohort.

In 1987, Madsen entered high school, and joined the Physics Team.

On a team field trip to a planetarium in Copenhagen, he was taken with the footage of the International Space Station broadcast on the planetarium's giant rounded immersive screen on 90-millimeter celluloid. He claimed that the experience left "a violent, almost physical impression" on him. He compared it to "a heroin trip." He was hooked.

Disillusioned with what he perceived to be his peers' lack of seriousness, Madsen quit school and decided to dedicate his focus to ocean exploration (without entirely giving up on his enthusiasm for rockets). He pursued an independent education, taking welding courses and apprenticing with various engineers, with the ultimate goal of fashioning his own submarine.

"Peter is unique," a former friend said, "but he is not a social being. He is an engineer. You can't have a normal conversation with Peter . . . You can ask him the same thing three times in a row and get three completely different answers."

Madsen convinced a foreman colleague to lend him workshop space rent-free, and began to design his first sub, living in that workshop amid machine parts, sleeping on the concrete floor. He became known as a quirky resident eccentric, and his submarine projects attracted the attention of the local newspapers. In 2008, Kristian von Bengtson saw one such article, which also mentioned Madsen's interest in rockets, and reached out to him. By this point, Madsen was no longer living on his workshop floor but had moved into an old scrapped ferry that had been decomposing on Copenhagen's south harbor. The quarters were rough, but at least he was able to wake up and fall asleep listening to the sea. There, living and working seaside, he constructed *UC3 Nautilus,* then the largest amateur submarine in the world, which he successfully launched in May 2008.

Von Bengtson went out to meet Madsen at the old ferry, and to gawk at *Nautilus.* "It was a dark evening," Von Bengtson remembered, "and I walked a little around this old, big ship to find him. He was in the middle of doing something, fiddling with something, which he then dropped when I came." Von Bengtson laid out his credentials to Madsen. He had worked as an aerospace architect for NASA and for the European Space Agency. He now wanted help in building his own rocket. On that rickety repurposed ferry, the two men spoke deep into the night and hatched their dreamy plans, collaborating on sketch after sketch of their space rocket ideas. Von Bengtson was taken with Madsen's fiery ambition as it pertained to both amateur submarines and rockets. "At times it was difficult to keep up with what Peter was doing," Von Bengtson said, "and the amount of thoughts that poured out of his head."

Later that year, they cofounded Copenhagen Suborbitals (which, as of this writing, remains operational), in Copenhagen's Refshaleøen district. They called it "probably the most ambitious private space program in world history." "The plan was actually very clear," Von Bengtson said. "We wanted to shoot Peter into space . . . That was it." They soon attracted a volunteer workforce of about fifty, which Von Bengtson described as "a geeky community of space enthusiasts." They created a website and solicited donations through crowdfunding. They held open houses that attracted hundreds of people, many of whom opened their wallets for these two burgeoning local celebrities. Over one thousand private donors contributed regularly to the endeavor, which Madsen also depended on to live. "It was the Paradise Hotel for engineers," Von Bengtson recalled. The two men enjoyed their honeymoon phase. Initially at least, Von Bengtson referred to their relationship as "love at first sight." Madsen, as was his nature, was fiercely dedicated to the new company, while making sure to leave himself enough time to engage his submarine enthusiasms.

Since 2011, Copenhagen Suborbitals has launched six of their rockets and space capsules into the air over the Baltic Sea, oftentimes using Madsen's home-built sub *Nautilus* as a launchpad—rockets named HEAT-1X and HEAT-2X, Nexo and Spica, Smaragd and Tycho, the latter named after the sixteenth-century Danish astronomer who, during a drunken duel with his third cousin over who was the better mathematician, famously had his nose cut off, and spent the rest of his life sporting a glued-on prosthetic made of silver and gold. As if Tycho wasn't enough of a sobriquet, the Suborbitals board lent the rocket the secondary nickname of Beautiful Betty after a donor who presumably bore a working nose of admirable size and shape. Though the launch of Tycho was considered a failed one, as it caught fire soon after liftoff, resulting in a "high-impact splash down," the board of amateur enthusiasts at Copenhagen Suborbitals remained optimistic. "It's a unique dream and we can't think of a more fun and exciting project," one said, than to one day "fly an amateur astronaut into space and safely back... *Because it's there.*"

Over the years (at least until the dissolution of their partnership and friendship in February 2014), at a series of pop-up art exhibitions, rave parties, and BDSM sex clubs in the gutted warehouses of the shipyard, Madsen and Von Bengtson, often seen together donning matching yellow Suborbitals safety vests, had been hatching their plans to garner sufficient public funding in order to send Madsen—and subsequently any interested and well-heeled member of the public—into outer space. Perhaps Madsen felt that by immersing himself in the cosmos, he could counterbalance his addiction to sinking to the depths of the Baltic, searching for the sea monk and solitude in the womb-like capsule of *UC3 Nautilus,* in which, both in the sea and on dry land amid the weeds of Refshaleøen's shipyard, he had now been living.

In fact, this very characteristic—choosing as his habitat the physical manifestation of one of his experimental "projects"—was in part responsible for Madsen's breakup with Von Bengtson. Madsen felt that his desire to actually *live* on *Nautilus,* while Von Bengtson chose to live not on one of his rocket ships but in an actual apartment with his spouse, Karla (a notable animation director), and two children, indicated that Von Bengtson's priorities were compromised. "I experience how Kristian comes in the morning and goes home in the evening," Madsen said. "Here I realize that Kristian is a normal person. He functions like normal people . . . I didn't find it fair that we weren't equally dedicated to the project."

"Peter will be surprised that someone could think of going at 3:30, because they had to pick up children from kindergarten," Von Bengtson said. "It was probably a matter of total dedication. That it was either/or. So, he sleeps out there under his lathe, he gets up, and that's what he does, and there's nothing to get in the way of that." Increasingly, Madsen appeared less in control of his emotions, often screaming at colleagues and throwing tools across a room. "I remember that my children were terrified of him," Von Bengtson said.

Madsen accused his partner of "disloyalty" and a lack of seriousness; he simply couldn't fathom that Von Bengtson would value his own family over their "lofty creative collaboration." Madsen's incessant tantrums, public and private, over the matter drove the exasperated Von Bengtson to quit Copenhagen Suborbitals and dedicate himself instead to his family, as well as to his now-independent intention to "reach space and outside our solar system before I die." The shaved-headed Von Bengtson also began to shun the public wearing of his old yellow safety vest, replacing it with his ubiquitous blue-and-white Adidas zip-up jogging suit.

Von Bengtson's departure did little to calm Madsen's workplace tirades, during which he explosively derided his Copenhagen Suborbitals colleagues for their lack of vision and single-minded devotion, and so, in the summer of 2014, four months after his cofounder's exit, Madsen was forced out of the company by the remaining members. Madsen, wounded, incredulous, and hell-bent on destroying his former colleagues, formed his rival one-man company, Rocket Madsen Spacelab, the mission of which was, in part, to declare "war" on Copenhagen Suborbitals.

Madsen doubled down on his "immersion" into his projects, both literally and figuratively, and spent even more time in his own laboratory. He continued to live and fund his projects on the donations of monied benefactors who were seduced by his charisma. He amassed a cultish following of unpaid volunteers. When he wasn't actively working, and when said work didn't demand the spaciousness of his lab, he rarely left his "home"—the confines of his submersible. It took on a sour smell.

Though *Nautilus* was then the world's largest privately built submarine (which Madsen constructed over a period of three years for about $200,000 in donations, and which he called "the ultimate art project... A political message about individual freedom"), it was, as a living space, dark and cramped.

When arising from it, Madsen often emerged stooped and scowling against the daylight. His features were naturally squashed, with little space separating his narrow eyes, his pug nose, the twisted rubber band of his mouth, and the ways in which he had to contour his body and his features when climbing from his sub into the fresh air only italicized his resemblance to what the journalist May Jeong called "a toy troll," and what the documentary filmmaker (of the Madsen-inspired feature *My Private Submarine*) called a "modern-day Clumsy Hans." Madsen clearly felt more at home in the shadows of *Nautilus,* wearing the same soft coveralls, the same broken-in work boots day in and day out, away from the judgmental eyes of the surface dwellers.

# 3

KIM WALL, who by age twenty-seven spoke eight languages, was an EU diplomat stationed in New Delhi before she was a journalist. And before that, she was a high school student interested in international relations, already meditating on such things as climate change and nuclear weapons testing. And before that, she was a little girl growing up in the small ferry town of Trelleborg—so named for the stabilizing leaning poles that allow the famed local medieval castle, once a Viking stronghold (now in ruins), to remain upright. Trelleborg is the southernmost town in Sweden, at the bottom of the bottom of the country. The facades of the wooden rowhouses are sea-battered and salt-scrubbed, their once bright beams now pocked and peeling. Caking the stone streets and flower petals in the window box gardens are rice-grain slivers of faded yellow, blue, and pink paint. The air is fresh and saline, and the town seems preserved, embalmed.

Beyond the town's confines, expanses of wheat fields, rapeseed, oil plant, and sugar beet farms stretch and roll and feed the country. When the local crosswinds stir such crops, they dance as they would no place else—lithely, hypnotically, their stalks undulating, their leaves shuddering together and apart like television static. It's easy to believe that one may have been transported to the bottom of what was once a great ocean, the ancient seaweed having adapted to become these crops. It's easy to convince oneself that some bygone species of fish might manifest here midair and swim among the feathery wheat-tops, take cover beneath the beet greens.

Kim shared her birthday, March 23, with the anniversary of Patrick Henry's famed "Give me liberty, or give me death" speech; of Tsar Paul I of Russia's death by sword, strangulation, and trampling in his bedroom at St. Michael's Castle; the start of Lewis and Clark's grueling journey home after all their exploring; of NASA's launch of Gemini 3, the United States' first two-man spaceflight; and with the birthdays of Emmy Noether (Jewish German-American physicist and Albert Einstein's muse), Bette Nesmith Graham (American inventor of Liquid Paper writing correction fluid), Joan Crawford, and Chaka Khan. On the day Kim was born in 1987, Olev Roomet, the world's only remaining player of the ancient Estonian bagpipe, the torupill, died. A kind of music came in; another music went out.

Kim's parents, Ingrid and Joachim, nicknamed her Mumlan ("Snuggles") due to her affectionate nature. Kim was sweet with the family Saint Bernard, despite the rotten smell of the dog's coat, rank with sea salt, sand, the stink of dead fish. On the outskirts of town, she and her younger brother, Tom, would run through the fields of canola flowers to the shore, so they could watch the ferries dock after having traveled to their small town from Germany, Poland, and Lithuania. They counted ferries named *Copernicus,* and ferries named *Tom Sawyer,* ferries named *Spirit* and *Vision, Saga* and *Viking.* Kim, in her excitement, would often race ahead of her little brother, but she would eventually slow down, wait for the whimpering Tom to catch up. She would stop, and close her eyes, and feel the yellow petals of the flowers dance against her face, smell the diesel and herring in the air. She would breathe and breathe and breathe.

When Tom caught up, tugged at her sleeve, and snapped her from her reverie, they would begin running together again, toward the harbor's entrance road, which was lined with palm trees, meant to commemorate Trelleborg's position as "southernmost in Sweden." The town employs a seasonal grounds crew whose job involves uprooting the palm trees in winter and replanting them in a temperature-controlled warehouse, saving them for the following summer. Kim and Tom loved the way the cool wind aroused the fronds, compelled them to collide with one another like castanets. They stopped harborside to admire the statue of the nude woman—a likeness of the actor Uma Thurman's grandmother (who was raised in Trelleborg)—erected in 1930 to commemorate the "embrace" of the land and the sea. The statue stands proud and provocative, her arms extended like wings, her head ecstatically—mutinously—thrown back. She looks as if she's about to take off or be beamed up—about to leave this earth, and her tether to the corporeal, terrestrial. Kim and Tom stared at her and felt the stirrings of things to which they couldn't yet attach language.

Because Trelleborg is one of the few places on earth that does not have its temperature recorded by an official weather station, when Kim and Tom would point and whoop at the docking ferries, their skin was awash in mysterious temperatures. As the boats opened their holds, the smells of goods from so many parts of the world commingled with the leavings of the onboard cafeteria specials—usually goulash and bratwurst.

Here on the Trelleborg dock, in August 1917, during World War I, a seasick and exiled Vladimir Lenin made a brief landfall on his journey back to Russia from Germany on the *Queen Victoria*, hatching his plans to overthrow the pro-war contingent and inspire peace at the eastern front, and a subsequent "dictatorship of the proletariat." It is another testament to Lenin's often misguided fortitude that the odd olfactory cocktail of fish and flowers here compelled him to clench his fists, dab his face with his handkerchief, and successfully hold back his vomit, extending his nausea all the way home. Kim's parents were history buffs, so she and Tom likely would have known of this story.

Their parents were globe-trotting journalists with eyes for social justice—Ingrid, the writer; Joachim, the photographer. In photos, Ingrid's white hair often seems to be lit from within, her forehead crimped, squinting against the sun. Joachim has long graying hair, ample eyebrows, his bushy goatee encircling a crooked grin. Joachim took a picture of Kim on her thirtieth birthday. In the photo, she holds a glass of wine, and Joachim later commented on the image—on the years passed since he first had to care for Kim as an infant, on the wonderful journeys his daughter had made.

Ingrid claimed that when she was pregnant with Kim, the baby first began to kick during a long layover in New York's JFK Airport, as she was in transit from an assignment in the Caribbean back home to Sweden. It was one year after the unsolved assassination of Olof Palme, once the most progressive prime minister in Sweden's history, whose death represented to many liberal Swedes (like Ingrid and Joachim) the moment "that the country lost its innocence." Ingrid sat in the hard seat at her gate, stared out the windows at the planes taking off and landing, at the ground crew hauling suitcases and waving iridescent orange wands. She moved her hands over her belly, and she felt Kim there.

"To stop traveling just because we'd become parents was not an option," Ingrid said. And like this, Kim first declared her presence in the world in a liminal space, amid announcements about arrivals, departures, safety, delays; heat-lamp pizza, flaccid fries, burnt coffee, and everything bagels. She arrived in a way station, floating, while so many travelers ate their lunches from their laps.

# 4

BEFORE THE SUBMARINE, there was the diving bell, and before the diving bell, there was the ephemeral but persistent human dream of sinking, the nagging drive to embed the body into a balloon-like enclosure and float alongside the fishes. To see as they see. The diving bell—a rigid, airtight capsule lowered to depth and raised by a winch-driven cable from a support platform at the surface—was first chronicled in the fourth century BC by Aristotle, who wrote, "They enable the divers to respire equally well by letting down a cauldron, for this does not fill with water, but retains the air, for it is forced straight down into the water." Aristotle, who is often cited as the "father of marine biology and biodiversity," spent at least five years of his life on the coast of Asia Minor, and there he may have used the diving bell to observe and first classify and name our sea creatures.

At depth, he became the first to distinguish between the "blooded" and the "bloodless" animals; the "soft-bodied" creatures (octopi and squid), the "soft-shelled" creatures (lobsters, shrimp, hermit crabs), and the "shell-skinned" creatures (bivalves, gastropods, sea urchins, sea stars). He began a list of miscellaneous creatures—singular beasts he couldn't shoehorn into one classification: sea cucumbers and anemones, isopods and jellyfish.

Aristotle grew obsessed with mapping the goings-on of the watery world. It seems he was addicted to the ornaments of the deep sea and incorporated his observations of the movements of transparent jellyfish, for instance, into his thoughts on the nature of memory and the soul. He wrote that dreams were akin to waking life "moving in a wave-like motion, as in a body of water."

Aristotle was able to see his obsessions from a scholarly remove, and to recognize that his drive to spend so much time at depth may have begotten a sort of madness ("No great mind has ever existed without a touch of madness," he snarkily wrote). Of course, Aristotle would have been aware of the treatise on "madness" written by Hippocrates, titled *On the Sacred Disease,* in which the ancient Greek physician described insanity as a "wet disease . . . ascribed to a wet condition of the brain involving excess movement."

Aristotle believed that people—men, specifically—who were excluded from positions of power and civic influence, or men who had recently endured disappointment or failure, often tend to remedy that failure first via a brief period of self-isolation (in, for instance, a workshop or a laboratory or a diving bell), and then by resorting to violence. He further mapped this social philosophy onto his opinions of stage drama. Any tragedy worth its salt, he believed, if it was to bear any resemblance to the actual human condition, must incorporate acts of violence: "the broken and dismembered bodies" of well-drawn tragic characters must "occasionally [be] brought onstage."

≈

In 343 BC, Aristotle was hired as a tutor to thirteen-year-old Alexander the Great, to whom he passed on his obsession with the diving bell. Legend has it that Alexander commissioned the construction of his own bell, which he affectionately named *Colimpha,* or "Swimmer." In numerous medieval-era images of Alexander, he appears ensconced in a submersible, floating among fantastical creatures. In European, Arabic, and Persian adaptations of the *Alexander Romance* (a text on Alexander's life, fusing history, biography, myth, and legend-making originally drafted sometime before AD 338), the conqueror is likewise shown suspended in a capsule beneath the sea.

According to this bumping-and-grinding of history and legend, Alexander used his diving bell not only as a tool by which to explore and subsequently dominate much of the Mediterranean but also as a quiet retreat, an isolated space in which to meditate on the blood he had shed on his journey toward "greatness," and the blood he still had to shed in order to maintain and expand it. Thanks to the atmospheric influence of his diving bell, Alexander was able to conjure a sense of blissful calm, even as he was cutting someone's throat.

In 332 BC, when he was only twenty-four years old, Alexander besieged the city of Tyre by first assessing its harbor underwater in his diving bell. He then instructed a portion of his army to remove the obstacles he found there, allowing his troops to arrive by sea unimpeded. Emerging from his submersible, he and his army defeated the Tyrian army and overtook the city, where Alexander subsequently directed the torture and massacre of over eight thousand unarmed Tyrian civilians and the enslavement of another thirty thousand women and children. As he boarded *Colimpha*, triumphant, he ordered the crucifixion of another two thousand residents, right there on the beach. Legend has it he watched these executions with a serene expression, and finally closed the hatch of his diving bell and set sail only after the last of the captured had gone slack on the cross.

By the age of thirty in 326 BC, Alexander used this meditative murderousness to amass his empire, which included much of Africa and India. When he died two and a half years later on June 10, 323 BC (purportedly due to assassination by poisoning), his body was placed in a golden anthropoid sarcophagus, which was filled with honey. In this way, his body never actually touched the solid frame of the casket, but was allowed to float "happy and unvanquishable forever" (according to the seer Aristander) in the sweet thickness, a vessel itself, an arrested submersible.

≈

According to the neuroscientist Dr. Ali Venosa, "the human body is either one of the most vulnerable things on the planet, or one of the most resilient. It's true we can do amazing things—heal where we once were bleeding, attack and destroy unfriendly microbial invaders, even knit our own bones back together."

"But what exactly can we take?" Venosa is compelled to ask. "What are the limits of our survival, and what happens to our body if we cross them?"

At depth, the lungs contract, and the brain and heart can grow saturated with blood, unnaturally powered to the point of exhaustion, which can result in a pressure-borne vertigo-cum-euphoria, the sense of existing outside of time and space, but still ensconced in and "mothered" by diving suit or submersible.

But if one goes too far when scuba diving, or is aboard a submersible that fails at depth, one can die from bleeding into the lungs, as the organs, according to the Centers for Disease Control, "exceed the elastic limit of the lung tissue." Many things can go wrong, in fact; a scuba diver or submariner attempting an escape from a malfunctioning sub can be knocked out by the pressure, or have a seizure, or succumb to nitrogen narcosis or oxygen toxicity, or drown.

According to the biologist Dr. Neosha S. Kashef, if one is fortunate enough to resurface from such depths, the person "will appear inflated . . . eyes bulging out of [their] sockets . . . as if blown up like a balloon."

Though the hull of a submersible can protect us from the more extreme effects, submariners remain at risk. "If a person in a deeply submerged submarine rapidly surfaces without exhaling during the ascent, sudden expansion of air trapped within the thorax can burst one or both lungs," stresses Dr. Michael F. Beers, professor of medicine at the University of Pennsylvania. Of course, said effects manifest at depth during submersible malfunctions as well.

At depth, in the human body, the nitrogen in the air we breathe diffuses into our blood and tissues in a higher concentration than on land. When we surface, this nitrogen can form microbubbles in our blood and our tissues, resulting in decompression sickness. In short, our insides effervesce, pop and fizz as if carbonated. People often experience this trauma as a cocktail of emotional ecstasy and physical pain. Rhapsody jockeys with agony, beatitude with torment. "Behavioral and cognitive aspects of cerebral decompression sickness may be persistent or slow to improve," writes the neurologist Dr. Herbert B. Newton. The sickness "manifests as an alteration of mentation." Frenzy can ensue—woe, intoxication, ravishment, madness—the sort that can compel a person to laugh or howl in ways that may seem inappropriate or eccentric; to exhibit wild behavioral outbursts before crashing, exhausted; to experience life—actually and metaphorically, physically and emotionally—in rapidly alternating periods of decompression and recompression. It's the compression that's the common denominator. Can we ever fully recover the equilibrium we once had inside us before we indulged in our compulsion to sink?

~

Imagine Aristotle bobbing amid the ancient depths, fogging up the glass of his diving bell, watching a sea monk agitate the water. Imagine Alexander, a few years later, doing the same. What happens when ancient instances of madness sour—if only narratively—into genius? The men who manned the submersibles of old were not always the best of influences. But to their less scientific contemporaries, they were mermen, gods, larger than the life that imprisoned the mouth breathers on the surface.

And like our best gods, they were mad, and they were bloodthirsty, and they were celebrated, and they begat future men who confused greatness with meanness, and who used their advanced tools to make better bells. And these men sailed and bobbed like Aristotle and Alexander before them, and fogged up new glass, and spotted new creatures, and used their machines to arouse the curiosity of their contemporaries on whose bodies they took out their frustrations, and became, too, names uttered from so many lips—whispers, screams, totems of disgust and caution and malign curiosity. Once again, celebrity and murder found themselves conjoined, as the ocean floor seethed with metal and flesh, and down there, not a single sea monk opened its eyes. Because we now tell ourselves we know better; sea monks don't exist. But people do. For better or for worse, people do.

# 5

It's a beautiful day in Fairmont, British Columbia, an unincorporated village with a population of 476, famed for its hot springs, the natural mineral water seeping out of the soggy ground wherever concrete hasn't been laid. The hills are cast in a margarine light, and the lukewarm breeze is so gentle that one notices it only when their arm hairs settle. The roads spread out arachnid-like from the town center toward the thirteen downhill ski runs, the tube park, one strip mall, and a gift shop where one can buy fudge, candles, pewter owls, and lots of moose paraphernalia. The town is surrounded by streams in which salmon breed, and if one closes their eyes, one can almost hear it—fish body slapping against other fish body. Some of the homes, evacuated during the great Fairmont mudslide of 2012, haven't been reinhabited, and malignant blooms of multicolored mosses overtake them.

About five kilometers from the center are the Dutch Creek Hoodoos, a sixty-seven-acre conservation area that supports badgers, woodpeckers, and eagles. According to the Ktunaxa Nation's creation story, in ancestral times, before the appearance of humans, when animals lorded over the earth, a primordial sea monster called Yawunik slaughtered any animal that crossed its path. In response, the animals formed a council and elected as their chief Nalmuqcin, an animal so large it would bump its head on the sky if it stood, and so had to crawl.

After numerous failed attempts to trap Yawunik, the council decided that Nalmuqcin should lay his huge body across the Columbia River—forming a dam, and thereby capturing Yawunik in the lake. The plan was a success, and the animals killed their tormentor, dragged his remains ashore, skinned him, and scattered his rib bones, which grew over the years into the Dutch Creek Hoodoos. The earthbound had finally rid themselves from the curse of the water monster. Nalmuqcin, ecstatic at the outcome and wishing to celebrate the victory of the animals, forgot himself and stood to full height, hit his head on the sky, knocking himself dead. His body became the Rocky Mountains, and through the tear in the sky the animals slipped, and time passed, and humans came along, and those animals, from their lofty vantages, acted as spirit guides to the people.

Like the sea monk, Yawunik straddles the actual and mythological worlds. According to the *National Geographic* article "Scientists Uncover Yet Another Cambrian Weirdo," "the archaic arthropod is a real animal that undulated through Earth's seas 508 million years ago." Resembling "a Lovecraftian nightmare," the *Yawunik kootenayi* "adds to the wonderful and perplexing spread of body plans that had evolved by this chapter in Earth's history—jutting out from beneath the invertebrate's tough exoskeletal hood are paired, pinching appendages arrayed with long wisps. The overall effect is of a lobster tail that's out for revenge on those who drew butter against it." A recently discovered fossil finally dragged it from the realm of legend to our real world—with its real rocks and real water. It lived here, and it ate here, and it mated here. It killed and it died here.

Adjacent to this ancient site whereon Yawunik was vanquished, nestled among tall, skinny, brush-topped trees, is Hank Pronk's workshop. Inside, Pronk stands in loose-fitting jeans and a black long-sleeved shirt. The windows of the garage door have been scrubbed translucent by the minerals in the air. Still, the day is bright, and light pours through them, illuminating the faded red paint of *Elementary 3000*, Pronk's pride and joy—the deepest-diving homemade submarine in the world. Pronk can't stop touching the sub, putting his arm over its flank as if over the shoulders of an old friend. Whenever he's near the thing, he can't contain his enthusiasm, though he tries. Hank is downright fatherly when he says he's "spent three or four years" working on the thing. I have traveled all this way to speak to him, desperate to uncover further, ideally less ominous, "answers," or at least inspirations, behind the compulsion to sink to depth.

Though Hank Pronk bills himself as a "very private person," his delight frees itself, and his deep dimples grow deeper, when speaking about the vessel. Occasionally he lifts his right hand—the one not resting on the sub—and brings it to his mouth, trying to conceal his smile. He's about six feet tall and sixty years old, doughy but sturdy. He has big hands. His hair is short and neat and graying, and his eyebrows are generous and expressive. They slither as he speaks.

"I built it to be extremely safe," Pronk divulges, unprompted from the get-go, as if anticipating a question about the dangers of sinking to depth in a homemade submersible. He's proud of his sub's safety factor, proud that some of the more innovative safety features were "improvised." "It was important to me to test myself," Pronk says, "and I wanted to build the deepest-diving homemade submarine in the world. Mission accomplished. This submarine dives three thousand feet." Pronk gestures with his head toward the sub and actually musses the top of it, as if the hair of a child who just won some race.

"Building a sub to a thousand feet is fairly basic," he says, "but going to three thousand feet is very, very serious."

Pronk is self-employed as a house mover, and nothing of his career experience prepared him to build submarines. He had to learn all the engineering techniques by watching online videos. "Just like looking up how to fix a leaking faucet," he says, "Google can show you how to build a personal sub." He doesn't claim to be a professional, but he's uncomfortable with being labeled as a mere hobbyist. "I've been building submarines since I was a teenager," he says. "Some people go snowmobiling, some people go dirt biking. I go submarining . . . Who doesn't want a submarine? There's only one other toy you could have that could be as cool as a submarine and that'd be a private helicopter."

Pronk circles *Elementary* like a sheepdog. "It'll be tested to over four thousand feet," he says, "and the crush depth is over five thousand feet, so it's a very solid and safe submarine."

The crush depth, of course, is the depth at which the submarine would implode due to water pressure. Many submarines and their crews, amateur and otherwise, have met their ends due to miscalculations of the crush depth. According to the U.S. Department of Defense (as of February 2016), a vessel's crush depth is typically determined by prediction "as to what a submarine's crush depth *might* be, [and] that prediction may subsequently be mistaken for the actual crush depth of the submarine."

Of course, this works the other way too, when pleasantly surprised crewmembers on board a seemingly doomed and sinking submarine find their bodies still delightfully intact when the vessel doesn't implode. In addition to *prediction* and *might,* the process behind calculating a submarine's crush depth bumps up against other contextually disturbing words in the Department of Defense's monograph, including *mistaken, misunderstanding, errors in translation, erroneous accounts, failure, evidently incorrect estimate,* and *general confusion.* Even the infamous and feared World War II–era German U-boats often imploded at depths of only 660 to 920 feet, which makes Pronk's backyard contraption positively miraculous by comparison. Still, beholding it— smaller than one might expect (Pronk himself stands well over a head taller than *Elementary 3000*)—one may feel less than eager to descend to five thousand feet in the thing, trapped in an enclosure not much more expansive than that of one plastic backyard kiddie pool downturned atop another.

The more one watches Hank Pronk, the more one may get the sense that he has carefully rehearsed everything he is saying—at least as it pertains to *Elementary 3000.* Before speaking, he'll take a sip of air and turn his gaze to the ceiling or sky, his thick eyebrows arching, as if remembering his lines. Then he'll deliver them in a modulated way as if holding something back.

For a guy with such an unbridled passion for sinking to depth that he would go to the lengths of fashioning the world's deepest-diving homemade submarine in his garage, his manner is decidedly bridled. He admits that as time has passed, he finds himself wanting to dive more often. He admits he's "kind of bored at home."

Yet when I ask him, albeit too excitedly, "Some DIY submersible builders have described the sinking as becoming an addiction. Would you characterize it in that way?" Pronk reacts like a deer in headlights and refuses to answer the question. Borrowing from his demeanor, I return the conversation to *Elementary* itself. I listen to him gush about the size of his sub.

"The size . . . ," he muses. "That's a biggie for me. My last submarine I just sold weighed six thousand pounds. Fantastic submarine, but a real bugger to handle being so heavy... So this submarine had to be under twelve feet and three thousand pounds maximum. Again," Pronk says as forcefully as he may dare, "mission accomplished."

He pauses as if to let the impact of the line—and the accomplishment itself—sink in. His jeans are sagging, his black leather belt not doing its job. When Pronk squats to point out the dome (or what I would have called the porthole), it appears as if his big knees are about to pop through his jeans, the denim worn thin and white. He caresses his sub's underbelly. The bottom section of *Elementary* is painted white, and the dome is beautiful—a convex lens of frosty glass set into a silver circular frame, symmetrically studded with golden bolts. It looks like a piece of jewelry; a giant earring.

"I cannot express to you how proud I am of this dome," Pronk gushes. "I made this dome from scratch." He says he needed to invent an industrial annealing oven, which is used to heat metal and glass at high temperatures to reduce hardness and increase ductility. "[I had] a controlled system that logged the data, and I could literally be [away], look at my phone, and tell what the temperature of the oven was."

Pronk runs his fingers from the dome, his pinky lingering on one of its golden buttons, down to the distended white belly of the thing— the occupant's sphere, or, as Pronk intimately calls it, "this sphere that I sit in." It is a one-inch-thick oversize coffee cup of pressure-vessel steel. It's formed of "two hemispheres," Pronk says, "custom bent within the required sphericity. That's the key here obviously." And I too want to live in a world wherein sphericity is the key.

Still squatting, Pronk inventories the submarine's parts—the pressure bladder, the event hole, the detachable drill he's filled with oil, the arm. It feels like an anatomy lesson, a dissection. The arm looks like a human arm made of metal, bent at the elbow, and cradling the bottom vertex of the sub's belly. The arm is an agent of exploration and—like a real arm—touch. Though it looks menacing and viselike, the arm's claw is apparently calibrated to be delicate (though, according to the Wyss Institute for Biologically Inspired Engineering, such arms often tend to be "hard and jerky and lack[ing] the finesse to be able to reach and interact with creatures like jellyfish and octopuses without damaging them"). Unlike the human arm, the Wyss Institute stresses, "the robotic 'arms' on underwater research submarines" aren't made to "turn a key in a lock [or] gently stroke a puppy's fur."

Pronk shakes his head. "I can pick up a banana and not leave a mark on it," he says, and moves to a part of the submarine resembling a miniature vacuum cleaner with a clear cylindrical chamber he affectionately calls the "sample sucker." With it, Pronk says, "We suck up . . . the bottom [of the lake bed]—sediment, weeds, whatever—so scientists can analyze it."

The arm and sample sucker are attached to a hinged base that is built to have a little give, since, Pronk admits, "I'm a bit of a wild pilot and I run into things, and I hit the bottom and whatnot . . . [and] a submarine of this nature with such a small occupant's sphere has a very small amount of volume; air volume." Clearly, Pronk is aware of the disasters that have long plagued, and continue to occasionally plague, the DIY submersible community. To help mitigate these obvious dangers, he has mounted "two huge carbon fiber [air] tanks . . . to add buoyancy to it, just to make it stay on the surface." The tanks are bright orange, and one has a sticker on the side reading INNERSPACE.

Anchored to the sub's flank is a black adornment resembling a rocket or a missile, with a little sharklike dorsal fin at its center, and a miniature propeller at its tail. Pronk grabs it from the propeller end and smiles before forcing the smile away. His eyebrows are vibrating with the effort. "This," he says, "is my own personal invention. I don't know if anybody's done it . . . These motors are not jettisoning normally. So if they get caught in a fishing line or a rope or anything like that, I'm stuck on the bottom. So what I did: I mounted 'em"—he grunts and yanks the little black missile-shark off the side of the sub—"on *magnets*. So these motors will tear away when they are tangled . . . I'm hoping it works well. Stay tuned, I guess.

"Now," he says, and pops up to full height, spanking *Elementary*'s side, "the body. This is pretty funny, actually. I didn't want to spend three or four thousand dollars laying up a body on mold. Blah-blah-blah. I hate fiberglass. So I went on Facebook . . . and I asked for people to donate canoes, bathtubs . . . And what I did was I cut them all up into pieces . . . fiberglassed them together, and that's how I made the whole body."

Pronk quickly gets serious again. "As soon as the ground thaws outside," he says, "I'm gonna dig another test pool beside the shop, line it with a pool liner, put a gantry crane above it, and start testing. Once that's all done, we'll start diving the submarine in Kootenay Lake . . . I do a lot of diving in the winter because the water clarity is the best."

His eyes become glassy, and he stares off into space, pining. "You can sit there all day no problem . . . ," he says wistfully. "I just take the sub out by myself. I cruise out into the lake and I dive . . . Often I'll just sit on the bottom of the lake and watch the fish go by . . . and eat my lunch. Kootenay Lake doesn't freeze, so that's where I go in the winter. We're gonna look for this lost gold boulder."

The gold boulder in question is the fixture of a local legend, a cautionary tale about greed and trespass, and, according to a local historian, "full of intrigue, betrayal, and murder." Though many residents today question the veracity of the legend, historians have uncovered newspaper articles from the 1890s testifying to the boulder's existence, including one published in *The Nelson Tribune* on April 23, 1894.

In the 1890s, a lucky prospector uncovered this gold boulder impacted above Kootenay Lake and, in exchange for a percentage of the loot, enlisted the help of a youthful miner with a reputation for stamina. What the prospector didn't yet know was that the miner was having an affair with his wife. The two men devised a plan to lower the gold boulder to the shores of Kootenay Lake, where the prospector had a boat moored, using a series of ersatz pulleys and scaffolding. The project took longer than expected, and the more their bodies ached, the more the men drank deep into the night. Lips loosened by alcohol, the prospector and the miner said things they shouldn't have said. Some believe the ensuing argument was over the division of the fortune. Some believe the miner let slip that he had been sleeping with the prospector's wife. The two men had among their cache of supplies a single rifle. They both raced for it, but the miner—who was younger and stronger than the prospector—grabbed it first and shot the prospector in the head, killing him.

Heart racing, the young miner concocted a story about an accident and planned to convince his mistress—the prospector's freshly widowed wife—to flee the settlement with him. But first, the miner trudged uphill to reckon with the gold boulder. Drunk, he fashioned a lever from hardwood, popped the boulder from its berth, and let gravity do the rest. Of course, once a boulder starts rolling, it doesn't decide to simply stop at a lakeshore. The great gold boulder rolled down the mountainside, splintering trees, splashed into Kootenay Lake, and sank.

Mortified and empty-handed, the miner returned to his mistress, who noticed the bloodstains on his shirtsleeves and detected the holes in his story. She dismissed him, and, dejected, he fled into the mountains. She reported what she correctly deduced to be her husband's murder to the local authorities, and they tracked the miner to the spot where the gold boulder once lay and executed him there by hanging.

Some say the prospector's wife became a proficient prospector herself—her life and her luck protected by the power of the gold boulder at the bottom of Kootenay Lake. Some say that, today, the gold boulder retains its power, protecting the well-being of the prospector's wife's spirit in the afterlife, cursing all rapacious men who attempt to search for the legendary rock, to get their greedy hands on it.

As such, Kootenay Lake has a dangerous reputation. A disproportionate number of people who dive into it—in a submarine or otherwise—don't return alive, inspiring such headlines as "Sudden Drowning Death in Waters of Kootenay Lake"; "Remembering My Brother and the Six Others Who Drowned in Kootenay Lake, 57 Years Later"; and "Dangerous Oasis: The Fatal History of a Popular Kootenay Lake Beach."

One article claims that "locals have feared [the lake] for decades."

"For many years I was afraid of Kootenay Lake," another local confirms, "afraid of all water, afraid that like my brother I would be swallowed up and never, ever found . . . and I am drawn to Kootenay Lake, the elusive yet beautiful body of water that holds my brother's body."

Even strong swimmers testify that the water here has an uncommon power, as they "felt the lake pull them away from the shore." "I can remember coming out of that lake," another says, "and I couldn't stand up. The headache was unbearable. I was basically dragging myself out." And yet another says, "[It] starts to just suck you in and you can't get out because it's pulling on your legs . . . [And the sand] sucks you down. And that's why they drown. They get sucked down and then people trying to rescue them can't find them."

Folks who have narrowly escaped the lake's pull, as well as those who have powerlessly borne witness to those who have gone down, oftentimes never return to Kootenay Lake. It's a place echoing with screams—those current and those remembered.

According to FortisBC, which operates four dams in the area, "the west arm of Kootenay Lake . . . is also more like a river, with a current that runs west and can drag swimmers out into the middle of the lake even when the water appears still."

The bodies of most of the drowned have never been recovered, and they remain, with the gold boulder, at the bottom of the deep lake. There have been so many that a monument has been erected, listing, in row after row, the names of the dead. Warning signs ornament the tree trunks, many rendered nearly illegible beneath graffiti. Residents have taken to making and posting their own warning signs along the trails to the lake, but someone keeps coming in the night and tearing them down.

~

When asked why, in spite of the dangers, Pronk still wants to engage in what is assured to be a folly, he says resolutely, "It's a fun thing to go and do."

Certainly, Pronk's heard the stories, knows about all the fatalities. I envision him down there, *Elementary 3000* battered by the under-lake currents and blinding storms of silt, its little magnets still holding fast. I envision him down there, amid corpses. Though it seems as if he's suffering my questions, I like Hank Pronk. And it's this realization and my worry for him that causes me to momentarily lose myself and ask him personal questions he will dismiss as too probing, too intimate. I break the rules. I use words like *obsession* and *compulsion*, *addiction* and *childhood*, *detriment* and *regret*.

One after the other, I ask him, "Has your compulsion to sink to depth affected other areas of your life?" "Can you talk about the origins of your desire to sink to depth in water? Did it occur in childhood? Did it have anything to do with the golden boulder?" Pronk stiffens and keeps stiffening. His eyebrows come to rest, as if at a lake bottom. His lips fall. He sets his jaw. And I know it: He has shut down. He holds up his right hand. He stops me.

"I think I will pass," he says. "I am quite private," he says again, "and your questions scream drama." His body looks huge and immovable. He makes of himself a wall, a hull, a sphere protecting the privacy of its occupant.

"If you want to find someone who will talk to you, go find the PeaceSubs group. They're the gaggle that got me into this in the first place. Lots of these guys love to talk about their subs."

I close my mouth. "Thanks," Pronk says, and like that, the interview is over.

# 6

A DISPROPORTIONATE NUMBER OF "Navy and Marine Corps Personnel Killed and Injured in Selected Accidents and Other Incidents Not Directly the Result of Enemy Action" perished, according to the Naval History and Heritage Command, on submarines. On December 10, 1910, an engine explosion on the submarine USS *Grampus* claimed the life of Electrician 2nd Class Herman William Ley, who was a visiting guest on the *Grampus* from the crew of the USS *Fortune*. On January 9, 1915, an explosion in the firebox of the boiler of the submarine tender USS *Fulton* killed Fireman 1st Class William J. Flaherty. Just over two months later, on March 25, 1915, on a beautiful, clear, windless afternoon, the battery of the USS *F-4* malfunctioned off the Hawaiian coast, and the submarine sank far beyond its intended depth, and because it was unable to rise, all twenty-one crewmembers on board drowned.

The list goes on and on, year after year. In 1917: USS *A-7*, gasoline explosion, Manila Bay, seven crewmembers burnt to death before they could drown; also 1917: USS *F-1* collided with USS *F-3* off the coast of San Diego, nineteen drowned; 1918: submarine chaser *SC-209*, mistaken for an enemy sub and attacked by Coast Guard gunfire, eighteen drowned; 1919: USS *G-2*, mysteriously sank at its moorings off New London, Connecticut, and never rose, three drowned; 1920: USS *H-1*, run aground and subsequently sank off Santa Margarita Island, California, four drowned . . .

These incidents persist, oftentimes shrouded in elements of curse and enigma, and are so numerous that the timeline of these tragedies matches up with the timelines of other aspects of our histories. In other words, whenever something notable (or unnotable) has occurred, there is likely a corresponding submarine accident to converse with it, to match and shadow or undo and recontextualize (if only via proximity) the events that have otherwise attended our lives, shaped our stories—every birth, every death, every marriage, every divorce, our birthdays, our anniversaries, our holiday dinners, our weekly chores—undergirded, haunted, dampened, and perhaps even enlivened by this bizarro other-story, this trajectory of submersible disasters.

November 10, 1966: USS *Nautilus* collided with another sub "during maneuvers" off North Carolina—one killed. May 21, 2002, a Tuesday, 58 degrees, zero precipitation, wind speed 12 miles per hour, just before midnight: the USS *Dolphin* mysteriously caught fire and began to flood one hundred miles off the coast of San Diego. The crew of forty-three were injured but rescued, and many went on to suffer from "post-immersion syndrome," during which their vocal cords closed over their windpipes, and many endured long-term aftereffects: chronic headaches, a vacillation between periods of intense sleepiness (during which recurring drowning nightmares were common) and sleep*less*ness, chronic coughing and chest pain, increased susceptibility to bacterial pneumonia, and the exhibition of unusual, irritable (and sometimes violent) behavior. The wreck of the *Dolphin* was abandoned at the site as the stunned crewmembers were carted away. Not long before this accident, the *Dolphin*, helmed by some of these same crewmembers, had set the new record for submarine diving depth at over three thousand feet.

In August 2000, the Russian Navy's nuclear-powered Oscar II–class cruise missile submarine K-141 *Kursk* sank into the Barents Sea when a hydrogen peroxide leak initiated the explosions of seven torpedo warheads, flooding the vessel with a high-pressure gush of seawater, immediately killing ninety-five crewmembers. Another twenty-three sailors briefly survived, huddling together in the sub's stern for several days, after which, progressively weakened by the lack of oxygen, they died of suffocation.

In May 2003, the entire crew of seventy died aboard the Ming 361 sub off the coast of Liaoning in northeast China when the diesel engines malfunctioned and consumed all the oxygen meant for the lungs of the crew.

On August 29, 2003, again in the Barents Sea, the decommissioned Russian sub *K-159* sank as it was being towed to a scrapyard. Newspaper reports referred to those on board as a "skeleton crew," of which nine of the ten people died. The sub's last radio transmission came at 2:45 a.m., pleading for help: "We're flooding! Do something!" In the aftermath of the disaster, many wondered why any crewmembers at all were aboard *K-159*, which has been described by the sub's retired former commander as rusted through, "a leaky, dark barrel that weighs 3,500 tons," its outer hull "weak as foil." The vessel had been decaying since 1989, leashed to an isolated pier on a rock in Gremikha Bay nicknamed "the island of flying dogs," for its intense winds. Russian newspapers wondered if the month of August was "cursed."

The sole survivor of *K-159*'s sinking, Maxim Tsibulsky, was, for reasons never disclosed, kept isolated from the public, from curious journalists, and even from his fiancée, Nadia, by a "double ring of guards" working for the Russian government as he convalesced in a noxious hospital bed in Severomorsk, a closed military city. Anyone outside the Russian military had been banned from asking any questions about the accident. The Russian government begrudgingly agreed to pay the equivalent of $330 in compensation to the nine victims' families. President Vladimir Putin's sole comment on the incident amounted to "The sea demands discipline and does not forgive any blunders."

The Russian government soon thereafter also banned the towing of any decommissioned nuclear sub, of which there were at the time at least a hundred we know of, each with nuclear reactors rotting into the sea, boosting the radiation levels of the water and the fish therein. The area in question happens to be the busiest cod fishery in the world and supports some of the largest habitats of haddock, and populations of king crab and polar bears, walruses and whales, all now enriched with uranium. For the time being, these subs, however volatile, would stay where they were, prompting *The Barents Observer*, a Norwegian newspaper, to dub the problem "Russia's slow-motion Chernobyl at sea," the vessels beholden to what then–Russian defense minister Sergei

Ivanov called "this frivolous Russian reliance on chance." Outside of Ivanov (who went on to address the cloistered Tsibulsky on television, saying, "There are no complaints against you"), many of the sources who later commented on *K-159*'s sinking and Tsibulsky's whereabouts were identified in reports only as "unidentified."

"I managed to reach the hospital," Tsibulsky's fiancée, Nadia, said, "and was told . . . his condition was grave. They didn't allow me to talk with him."

"How Maxim Tsibulsky escaped from the sinking *K-159* nuclear submarine—and what he saw and heard that day—is no doubt an incredible story. But Moscow apparently doesn't want it to be told," claimed an article in *The Globe and Mail*. Tsibulsky was coerced into signing "a waiver promising not to tell his side of the story to the press," and was being interrogated as "as a witness in a potential criminal negligence case."

Nadia was banned from entering the city of Severomorsk. Citizens, hearing rumors of Tsibulsky's detention, sent bouquets of flowers to the military hospital. As the flowers were meant for a living person, they were grouped, per Russian superstition, in odd numbers (even-numbered bouquets are meant for the graves of the dead), but they were all discarded before reaching Tsibulsky's room, tossed into the gloomy and sodden back alleys of Severomorsk, where they were marked by the stray dogs. At one point, Tsibulsky tried to sneak a phone call to his father, Viktor. He spoke quickly, able to utter, "I'm fine, Dad, I'm healthy. Please call Nadia, I cannot reach her," before the guards severed the connection. What we do now know is that doctors considered Tsibulsky's survival to be an inexplicable "miracle." The maximum time a human can spend in water as cold as the Barents Sea and survive, the doctors determined, is forty-five minutes (though most people would perish long before that). Tsibulsky was immersed in the frigid waters for nearly two hours, and more than one doctor wondered as to the precise nature of his "humanity."

Tsibulsky, though, expressed very human emotions, telling his interrogators he was "terrified" while waiting to be rescued, and he "could not remember which of his crew mates had accompanied him through an escape hatch," only that they "died before my eyes." The winds were so strong that night that, maybe, dogs flew through the air. Tsibulsky's heart did things it shouldn't have been capable of. The fact of Tsibulsky's survival made the doctors and the government officials alike doubt both his humanness and his testimony. They saddled him with the stuff of legend; the stuff that begat superstition. He was alien or robot or god or merman. He was an odd number of flowers; a flying dog.

He was a power the military wanted to dissect and harness. One government official tried to explain it away, telling Tsibulsky, "You were wearing your sailor's shirt, that's what saved you." Some speculated that he had more air in his blood than a normal person, that he was, in effect, part balloon. Of course, no one in the room bought it. But we do know that the government offered him a nice wristwatch as compensation for his ordeal, plus the 330 bucks (handed to him in a damp envelope). And we know also that myriad Russian newspapers believed the story of the accident as perpetuated by the Russian Navy to be rife with lies—from the cause of the accident, to the condition of the sub, to the weather reports (the Navy claimed that a violent squall was responsible, but the newspapers reported, "There was no storm. Admirals again tell lies").

*K-159* reportedly sank to 240 meters, interring the corpses of the crew. It sits on a piece of seafloor close to the wreck of *Kursk*. The Russian Navy, for reasons inscrutable, also lied about the depth to which both of these ruined subs have sunk. In various reports, the distorted numbers remain inconsistent, sowing a sense of chaos via the cockeyed disinformation, but in each case, the depth reported was far shallower than the actual 240 meters, as if that larger number would make the Navy look worse, italicize and electrify their neglect, further expose the nerve endings of the tragedy. The responsibility for these deaths, and indeed the deaths themselves, these doctored numbers seem to tell us, can be more easily shrugged off if the submarine coffining the bodies were perceived to be in shallower water. In the greater, actual deep is reality and, therefore, reckoning.

≈

In spite of the many submarine disasters, and the sometimes violently delusional contextualization of them, more naval enlistees request to be on the crew of a submarine than any other vessel—be it sloop, frigate, patrol ship, cruiser, gunboat, schooner, steamer, destroyer, or battleship. When asked why, many are bemused by the question, unaware of any received cultural narratives or myths with regard to the submarine. Most shrug, wrinkle their foreheads, delightfully mystified by their own desires. Some use words like *adventure, romance.* One uses the word *renegade,* over and over. When asked if it has to do with the riddles of the deep, the actual measurable sinking to depth, many respond—quickly and excitedly, as if something spot-on has finally been pinpointed—with bravura certainty, in the affirmative: *Yes, yes, oh yeah, totally, for sure, 100 percent, that's it, definitely, definitely, definitely yes.*

One enlistee cites the author Jules Verne and his ambition, carried on from childhood, "to go down to twenty thousand leagues, you know?" Verne, in fact, would have been exasperated by this utterance, and may have accused the young man of a cursory read. Verne's iconic book—which is also cited by many amateur submersible builders as an early inspiration to them—was published in 1870, under the original French title *Vingt mille lieues sous les mers: Tour du monde sous-marin,* or *Twenty Thousand Leagues Under the Seas: A World Tour Underwater.* The distance of twenty thousand leagues (or fifty thousand miles)— which is nearly twice the circumference of the earth—refers in Verne's book to the cumulative distance traveled underwater across all the world's seas, and across all human history (up until 1870). The distance does not refer to any specific *depth* at all, let alone a depth reached by a singular vessel like Captain Nemo's *Nautilus,* because, as Verne well knew, this would be absurd. The average depth of the world's oceans comes nowhere close to fifty thousand miles; it is a comparatively mere (but actually awesome) 2.3 miles. The deepest part of the deepest ocean (the Pacific) is 6.7 miles, and the greatest depth attained by *Nautilus* in Verne's book (it's fiction, after all) is four leagues, or 9.94 miles, which is still rooted in exhilarating and otherworldly impossibility, but one that, in 1870 at least, bore the illusion of possibility, as opposed to something entirely ridiculous and dismissible as frivolous (like a depth of twenty thousand leagues, for instance).

This misconception among English speakers who either haven't read Verne's book, or who have read it while distracted, perhaps, by the overly fanned flames of their childhood daydreams, is the product of a mistranslation. When Verne's book was first published in English in November 1872, two years after the initial French publication, it appeared poorly translated by the London-based Reverend Lewis Page Mercier, who, in trying to hide his imperfect command of French (his Huguenot ancestry notwithstanding), cut nearly a quarter of Verne's original text, excised all references to Ireland's freedom fighters (many portraits of whom, in the original text, decorated the walls of Nemo's stateroom), scrubbed political and religious views interpreted as liberal (Mercier was a conservative British Protestant), infuriatingly translated the oft-occurring term *scaphandre* ("diving suit") as "corkjacket" (which, in old British slang, refers to a life jacket, sort of the opposite of a diving suit), and mistakenly confused the titular word *Seas* for the singular *Sea*.

Perhaps Mercier's half-assed translation was the result of his bitterness in having to take up translating as a side job in the first place, after he had accumulated a debt to a wealthy landowner. Perhaps his brain-space was better devoted to the feeding and rearing of his nine children, and to the composition of his weekly sermons in his role as chaplain at the Chapel of the Foundling Hospital, the congregation of which included such luminaries as William Hogarth, George Frideric Handel, and Charles Dickens. Anyhow, in spite of these glaring errors and offensive omissions with regard to Verne's tome, Mercier's translation (which he bragged took him less than four months to complete) served as the standard English-language version until 1966, nearly a hundred years after the book's original publication.

When Peter Madsen cracked open the book as a child, he was attracted to Nemo's dedication to living "outside of social laws, sailing the seven seas in search of total freedom," and likely to Nemo's near-violent breed of iconoclasm, as defined by the critic Margaret Drabble as "cultured, tragic, ruthless, wealthy, he is at war with humanity . . . a more complex figure than his three hostages." That Madsen's fixation on these qualities in Nemo all but drowned out for him Verne's obvious subtextual "criticisms of bloodthirstiness" (Verne tends "to deplore needless killing") and Nemo's "anti-imperial, anti-colonial . . .

compassionate" tendencies, is another act of cursory reading, misinterpretation, inadequate translation. Verne, in actuality, was skewering people like Madsen. *Nemo* is a riff on the Latin for "no one," and the name choice was meant to subtly underscore Verne's pessimistic outlook on the future of technology and exploration when those who undertook such ventures were driven by vengeance and megalomania. According to earlier drafts of the book, Nemo, a Polish scientist and nobleman, was "cast out of civilization" after being consumed by vengeance when his family was brutally murdered by Russian soldiers during the uprising of 1863–64. But Pierre-Jules Hetzel, his publisher, persuaded Verne to soften such indictments (of both vengeance and Russian politics) so as not to scare off a faction of his readership and alienate the Russian market.

Many critics claim that Hetzel was more Verne's censor than editor (with whom Verne argued passionately and often in vain to retain his authorial vision in a heated correspondence amounting to some 680 letters). "I am totally incapable of depicting what I don't feel," Verne wrote to Hetzel on May 17, 1869. "Obviously, I don't see Captain Nemo as you do . . . Nowhere, despite what your letter says, have I portrayed a man who kills for the sake of killing . . . His hatred of humanity is sufficiently explained . . . You are going to end up making me disgusted with this book." But Hetzel, and, by extension, the Russian market, had the final say. And Mercier, the blocky fundamentalist, softened and censored Verne even further via a mistranslation both deliberate and incidental, until Nemo became the sort of character who, according to one critic, "believes that innovation is the domain of a sort of self-nominated elite. Basically, he's a libertarian." When Madsen (along with many other submersible enthusiasts who credit the book for awakening their niche passions) picked up the book, it was easy to misinterpret Verne's intent and confuse indictment for romanticization.

As with those under the spell of a religious fervor who pick and choose the elements of, say, the Bible that already suit and justify their own feelings to the exclusion of other passages, so Madsen dressed himself in a limited set of Nemo's traits and built for himself (and subsequently moved into) a skewed version of Nemo's vessel, *Nautilus*.

That Madsen dreamed of sinking to the impossible and absurd depth as cited in Mercier's mistranslation is, of course, another mistranslation of reality, one that likely frustrated Madsen's delusional ambitions to be larger than life, and so he sought another avenue—one that would quash said frustrations, and allow him to sink, however metaphorically, to a depth of twenty thousand leagues.

# 1

THE NAUTILUS ITSELF—meaning the marine mollusk—hides away in its shell, which is the most perfect naturally occurring logarithmic spiral on the planet. This perfect spiral is so durable it can withstand anything the sea throws at it. But its perfection confounds us and frustrates us and compels us to try to "solve" it, understand its supple whorls, follow its chambers to some elusive end or bottom or depth. In this perfect spiral, we spiral out of control. Inside its shell are two "primary" chambers, each divided into sub-chambers. As nautili age, these sub-chambers proliferate and hatch additional ones. In one of the "primary" chambers, the little beast lives, arranging its sixty to ninety cirri tentacles, the soft appendages of which can retract into corresponding rigid sheaths. It has a powerful grip. An opportunistic predator, it rips apart its prey—mostly small crustaceans—with its parrotlike beak, and its muscular stomach acts as a crusher, pulverizing its food between its walls. Because it has such a short memory span, it quickly forgets what it has done. In its shell's second "primary" chamber is a cocktail of mysterious gases. Like submarines, nautili have a "crush depth" and will implode at about twenty-six hundred feet.

Its color variation makes it virtually invisible in the water. It is invisible in the light, and it is invisible in the dark. Like a ghost, it's there, but we can't see it. It haunts us. Paradoxically, the nautilus can grow and grow without ever changing shape or spilling its bounds. Nautili have existed in this form, unchanged, for hundreds of millions of years, and we are compelled to mythologize them, even though they are actual. We call them "living fossils," though they are not fossils at all. They live and love and kill and eat beyond our grasp and the grasp of our language. We fish for them and manipulate them, claim ownership and fashion them into art and service pieces—we make jewelry and goblets of their shells, make them the centerpieces of dioramas and curiosity cabinets. The composer Deems Taylor wrote a sad cantata about the nautilus. Oliver Wendell Holmes Sr. wrote a poem about it, comparing his own fleeting soul to "Leaving thine outgrown shell by life's unresting sea!" Andrew Wyeth painted a portrait of it, its shell entombing a woman laid out on a canopied bed. We etch nautilus shells with fanciful scenes of humans farming, and fighting, and making love beneath great migrations of birds and bats, amid packs of circling dogs, elliptical spiders. We inscribe them with miniature likenesses of eighteenth-century Royal Navy admirals named Horatio. We etch their perfect shells with less perfect renditions of the nautilus itself.

Each chamber in the nautilus shell's interior is mathematically beholden to an exact Fibonacci sequence—that confoundingly beautiful formula that attempts to map and explain the whorling things of the world. According to sacred geometry—which postulates that certain naturally occurring patterns are evidence of, or the embodiment of, a divine creator—the shell of the nautilus is the most robust evidence of the existence of gods and goddesses. As far as we know, Plato was the first to posit and popularize the idea of sacred geometry ("Plato said god geometrizes continually," Plutarch wrote). Plato spoke passionately of this idea to Aristotle—his student of over twenty years—who subsequently mapped notions of sacred geometry over and onto his early submersible designs, believing that the more "divinely" shaped his diving bell, the more the gods would protect him at depth. And, of course, Aristotle passed such ideas along to his own student, Alexander the Great, who used such a design formula in fashioning his own diving bells, the "perfect," "harmonious" whorls of which calmed his heart as he engaged in his bloodthirsty colonial campaigns.

The nautilus is our only earthly example of the golden spiral, mathematically governed by the golden ratio (1.6180339...), and if only we could traverse its depths without going mad in the face of such perfection, we would find, at its impossible end, an almost unbearable aesthetic perfection, the avatar of God.

*Nautilus,* in fact, was Madsen's third (and most advanced) home-built submarine. The first and second machines were named *Freya* (after the Norse goddess of love, and once the most popular baby girl's name in Denmark) and *Kraka* (after a Norse mythological figure known for her intelligence). Madsen launched *Freya* in 2002 and *Kraka* in 2005. According to the thirteenth-century legendary saga *Tale of Ragnar Lodbrok,* Kraka (who was then an unnamed toddler) was confined to the hollows of a massive harp built by her foster grandfather, Heimer, who worried for her security after the death of her parents. She was forced to contort her imprisoned body to match the tight spaces and curves of the instrument, as Heimer wandered the countryside with it, masquerading as a destitute freelance harpist.

Arriving at a tiny village in rural Norway, Heimer was invited to play his harp at the house of a local couple, Áke and Grima. They provided Heimer with their homemade moonshine, and together they drank, and the couple danced while the old man plucked his strings deep into the night. After Heimer fell asleep, leaning against his instrument, Áke told Grima he felt that a harp of such magnificent size must be hiding something within it—something of great value. Grima nodded excitedly and convinced Áke to murder the old musician while he slept. Áke complied, and with the notched blade of an ancient Viking sæx knife, he slit the old man's throat. As Heimer's blood decorated their walls, Áke and Grima took up their hammers and smashed the harp to pieces, discovering, to their surprise, the little girl inside. Immediately they were overcome with guilt, and as her foster grandfather bled out, they decided to raise her as their own.

Áke and Grima, though, were a homely couple, and were often ridiculed in their village for their distorted and mismatched features. Who would believe that this girl—who was uncommonly beautiful—was their own child? It was Grima who came up with the solution, and it was Áke who pinned the girl down as his wife poured tar over her body, tied her into a long, hooded cloak, and christened her Kraka, meaning "crow." They fed her exclusively on raw onions and so, imprisoned again, she retreated into a life of the mind and developed the sort of genius that would, many years later, endear her to the legendary king Ragnar Lodbrok. Eventually, the two would be married (only after Kraka denied his first few proposals, and refused him sex until they were wed, which, in legends of this sort, represented a heretofore unheard-of assertion, and therefore an early example of female empowerment and resolve). Today, the *UC2 Kraka* midget submarine, which took Peter Madsen two and a half years to build at a cost of 200,000 Danish kroner, rots on the grounds of Denmark's Technical Museum in Helsingør, its hull blooming with amoebas of rust and weeping ribbons of bird excrement.

After *UC2 Kraka* was decommissioned due to age and decreased functionality, Madsen began work on *UC3 Nautilus*. At a length of 17.76 meters, it was 5.5 meters longer than *Kraka* and completed at a cost of 1.5 million Danish kroner, likewise donated via crowdfunding sources. Its hull painted slate gray, and its rounded interior walls and ceiling painted puke green, the tight vessel, despite being larger than *Kraka*, still forced one to hunch when standing, the only light creeping in through the few scratched and salted portholes, their recycled panes further obscured by a slick film of old green sea scum. Every footstep into and within *Nautilus* elicited the dull gong of the metal, the hollow echo of vibrating steel. In this sound was something unnervingly final, the closing of a prison door. The vessel's two single cots, topped with crumpled blankets that one publication described as "wolf-gray," were separated only by a one-foot-wide walkway. In photos, the sheets on both beds appear rumpled and sallow, and littered with hoses and masks, backpacks and newspapers, toilet paper rolls and fire extinguishers, plastic bags, banana peels, apple cores, cherry pits, thick blue rope. In not one image is the place anything but a pigsty.

Beneath the bunks, Madsen wedged a series of metal drawers he solicited from defunct commercial airplanes in which, he said, "you can store various things safely." Beyond the pillow of one bunk was a microwave oven, and beyond the pillow of the other was a sink, on the edge of which, in one video, sat an abandoned plate, a tangle of spaghetti, the downturned fork still slick with Madsen's saliva, anchored into the noodles. Video game developers from Ubisoft re-created *Nautilus*'s interior design for their submarine game *Silent Hunter 5*. For *Nautilus*'s maiden voyage celebration on Saturday, May 3, 2008, the local installation art collective Half Machine organized four uniformly dressed ballet dancers to pirouette in time on the submarine's deck, their white-and-pink dresses billowing into the cool wind as they spun.

The launch celebration was a spectacle, beginning early in the morning amid the forklifts and cranes and dead grass at the soggy dry dock on Refshaleøen's west end, where *Nautilus* sat propped on concrete blocks. Blond boys ran through the puddles waving Danish flags and kicking the heads from dandelions. The day was bright, partly cloudy, and young mothers drew blankets over their strollers and carriages, protecting their fair-skinned infants from the sun. Still, the air was cool, and many wore red jackets, red wool caps. Crewmen anchored ladders against the side of the submarine and climbed them, affixing cameras and placards bearing the contact information of various sponsors. Some polished the hull with dishrags and spit. It took two men to affix the huge Danish flag to the submarine's conning tower, one of whom was Madsen himself, clad in a black Kecon A/S (Architectural and Structural Metals Manufacturing) jumpsuit, his blond hair wild and windswept, parading along the top of the vessel, delivering frantic orders to those below.

Men in wheelchairs, men holding boom mikes, women in knee-high leather boots, and leaping children all cheered as one of the cranes lifted the submarine from its moorings via three attached chains, six gaunt white windmills whipping in the background. The submarine twirled in the air, the submersible flying. Flashbulbs exploded. Cars and people backed up, getting out of the way. Someone tossed a yellow hard hat into the air. The spectators let out a collective sigh as the crane nestled *Nautilus* onto the oversize flatbed of a royal blue Rahbektransport semitruck. As the truck moved through the streets and smokestacks, winches and warehouses, the crowd grew and packed together. They opened picnic baskets and high-fived as a jazz band played, the yellow caution lights on the roof of the truck flashing. The Royal Danish Navy goose-stepped ceremoniously in front of the truck in their white caps, striped shirts, and ascots. People sat on other people's shoulders. People danced, and blew into saxophones, and rattled their tambourines.

Another crane lowered *Nautilus* into the Øresund Strait, and Madsen rose from within, climbed out onto the hull, and straddled the periscope. Onshore, people applauded, and Madsen paid them no mind. He straddled his hull and laid his hands onto the cool air. A skipper's hat sat squashed and askew on his head. It looked like a flattened loaf of white bread. His face in turns reddened and paled, reddened and paled. A photographer for a Danish newspaper referred to him as "a crazy person." A friend deemed him capable of operatic petulance, of often behaving as "a child who just lost his toy or dropped his ice cream." A Swedish newspaper photographer said of Madsen's endeavors, "Making space rockets and sailing around in homemade submarines is not normal behavior, but I've never seen him lay a hand on anyone." But former interns at his workshop cited instances of Madsen's wild tantrums, his habit of hurling screwdrivers and hammers at the heads of his volunteers if they made a mistake or said something that displeased him or exhibited behavior he considered to be too cautious and not worthy of his mission to make submarines, "to be free from authorities." Madsen himself dismissed such abuse, declaring, hey, "[they] all know they are taking part in a Peter Madsen project."

In spite of the abuse, Madsen and his unpaid crew would often reconcile by the day's end by participating in one of Madsen's odd skits in which he and interns would pretend to be Nazis. They recited lines and reenacted scenes from old Nazi propaganda films, Madsen correcting his crew's accents and their blocking. Madsen would give each their own "Nazi-inspired" moniker, and they would engage in various hijinks, headlocking one another and giggling, goose-stepping, saluting, miming historical executions, and making jokes about taking over the world, and exterminating the Jews, and stabbing one another in the kidneys, and injecting battery acid into one another's jugulars. Some days, amid the scattering of rocket and submarine parts, Madsen compelled his crew to speak only in German, and he referred to them collectively as the Third Reich.

Madsen spat into the strait. As the water undulated, his gaze became more and more focused. His mouth twisted like a little strand of rope.

Sæby, the small town in which Madsen grew up, is one hundred kilometers west of Copenhagen. Population 343, Sæby is the largest small town on the banks of Lake Tissø, the fourth-largest freshwater lake in Denmark. Tissø, via a complicated etymological journey, translates as the God's Lake, after Tir, the Old Norse god of war. Many villages in Denmark begin with *Ti*, designating the place as sacred, as fierce. Such designation does not exist in Swedish place-names, which thus provides some nationalistically minded Danes with an excuse to feel prideful, contemptuous of those "weaker" Swedes.

According to various Nordic folktales, the lake was created by a sow with eyes of fire and the teeth of a shark who, along with her nine evil piglets, released such an amount of excrement that the land buckled, and from the resulting sinkhole rose the cleansing waters of Tissø. Or the lake was created by a community of ten trolls who, annoyed with the persistent ringing of the Sæby church bell, each took up a rock and hurled it across the meadow and over the church's stone wall. But the bell, ever swinging, batted each rock back over the wall and into the meadow, and as the trolls ran for cover, the cumulative impact of the ten rocks opened up a hole in the earth from which lake water began to rise. Or the lake was created when the ten seals of the underworld, succumbing to their wanderlust, rose to the surface of the earth and poked their heads through, sniffed, dismissed what they saw as terrestrial mediocrity, and retreated to their depths. The collective divot made by their curious and scornful heads began to fill with water. Or the lake was created when the evil and incontinent sorceress Marie squatted with one foot on a rock named Big Bird, and the other on a rock named Little Bird, and filled the space between them with her gush of urine. Or the lake (its formation is often linked to the number 10) was created by a band of war-god-worshipping marauders to conceal the corpses of ten early Christian priests—tortured and sacrificed to Tir—which were buried in its depths.

Many believe this last explanation to possess a kernel of historical verisimilitude. The area surrounding the "holy" lake was an important settlement during the Viking age, and on its banks, still today, are the ruins of a Viking "slum complex" and nearby cult-house, in which have been excavated a disarming assortment of skeletons of various species—victims of both human and animal sacrifices to the war god during ritualistic feasts. The caved-in skulls of horses, men, women, and children have for years been heaved up by the marshland. Remnants of weapons (most of which date from the Viking age, but some dating back to the period before the year 0) have been excavated, waterlogged and scummy, from the lake—spikes and lances adorned with the rib bones of birds, the breastplates of infants. Teeth.

Runoff from the surrounding farms has turned the lake into one of Denmark's more polluted, and over twenty species of fish, all of which have miraculously adapted to the filth, thrive therein, spending their entire lives here, blessed by the war god with outsize fertility rates. Multiple generations of perch and pike will know nothing of this world beyond the confines of Tissø, its depths and shallows, meanness and holiness, crud and skulls. By law, these fish are subject to the threat of the hooks only during daylight hours, their meat contaminating the bellies and the blood of those who consume it. Some believe that the ingestion of the fish of Tissø leads, inevitably and torturously, to madness.

Peter Madsen grew up under the influence of water and war and bones. Of insomniac trolls. Of sorceress piss and the feces of demon-pigs. Overhead, he gazed at the largest concentration of birds of prey in Denmark, the eagles and the ospreys. He watched them hunt for the birds that spent most of their time on the lake's surface; watched them, meditatively, eviscerate the diving ducks and bean geese. From an early age, Madsen trudged along the marshy shores, flattening rare and threatened plants named sunrose, lousewort, helleborine. Madsen kicked the heads from orchids and pasque flowers (which were Isaac Newton's favorite). He was able to distinguish the different voices of the birds, and when the eagles descended, he was able to determine which of those voices had been stopped. He dreamed of deeper water.

His father, Carl—who was thirty-six years older than his mother, Annie—was a pub owner, and often came home from work drunk, red-faced, razor-burned, and pasty-mouthed, stinking of orris and rotten booze. An abusive man, Carl would regularly accuse Annie of having feelings for other men and beat her and Madsen's three half brothers (Annie's sons from two prior marriages), but spared his own progeny. Madsen was exempt from the physical manifestations of his father's furies and frustrations, the fist-shaped bruises that blossomed on his half brothers' backs, his mother's belly. Peter Madsen was special, clean, unreddened and unpurpled. He was his tyrant father's little prince. He watched the others recoil and whimper, but he did not feel their pain. When the abuse became overwhelming, Annie fled with her three older sons, but left six-year-old Peter behind with Carl, and soon, father and son bonded over their mutual fascination with rockets.

During World War II, Carl worked as a carpenter and helped to build the barracks for the German soldiers who were occupying Denmark. Carl developed an admiration for the Germans, and later told his son stories lionizing the Nazi aerospace engineer and rocketry expert Wernher von Braun. Father and son would fashion crude homemade rockets together and take them into the boggy lakeside fields. On the spongy ground, they would squat side by side, Peter's blocky shoulders miniature versions of his father's, and the waterbirds would pass overhead and ignore these two human-shaped shadows below, mistake them for harmless, until Carl, passing some incendiary torch to his son, took the book of matches from his pocket and slipped it into Peter's small hand. Peter was eager to please his father, but neither was adept at engaging the other emotionally. They often spoke to each other only of science—of things like gunpowder and hydrogen balloons and ballistic missiles. Peter struck the match, and the cool wind blew it out. Carl shifted some phlegm, cracked his knuckles.

"I wonder what Mom's doing," Peter said. Carl spat. A globule of saliva held to his lower lip. The daylight made it sparkle. Flatly, Carl said to the boy, "You can always visit your mother, but if you do, don't come back." Peter's back was sweaty. His neck was cold. He lit another match, and Carl cupped his hands around the flame until the boy brought it to the fuse, and the fuse hissed. And they stood, the two of them, Carl's knees cracking, and took a few steps back into the tallgrass. For safety. A couple of seconds later, the rocket ascended into the sky, and the waterbirds cried out in surprise and warning. And father and son looked at each other, feeling something akin to pride.

In the colder months, Carl and Peter would stay inside and binge-watch the 1981 film *Das Boot,* about the exploits of the crew of a German U-boat during World War II, and young Peter began to marry his fascinations of rockets and submarines. "The whole idea of sailing underwater," he would later say, "kept me dreaming for many years."

He dreamed at eighteen years old as he watched his father die, and he dreamed as he began and subsequently abandoned his pursuits of various internships—in welding, in refrigeration, in engineering. He spent these years living in the corners of various workshops, sleeping on mattresses on the floors amid greases and tools, amassing scant possessions—a few items of clothing, books on World War II, books on the components comprising rocket fuel. He joined the Danish Amateur Rocket Club but was soon excommunicated by his peers for what they felt were his "wilder and wilder ideas," and for his using "controversial" rocket fuels the other members deemed too volatile and dangerous. According to a former colleague, the other members of the club "see the rockets as a pleasant leisure project. But Peter does not have leisure projects. He only has projects, and they take up everything." The other members "ended up hating him. As they said back then in the club: If you say [Peter Madsen's] name, the sprinkler system starts."

To interrupt his solitude, and further frost what seemed to be a single-mindedness, Madsen began to surf various internet hookup sites, including Travelgirls.com (where he could find "thousands of adventurous girls who want to travel") and frequent BDSM clubs and private fetish parties, though often more as a spectator than a participant. He would sometimes show up in a naval uniform and cap. He tended to have one steady open relationship, and to seek out what he called a web of "crazy ladies" on the side. He befriended a woman who had broken both of her hands, and ritualistically visited her in her apartment every day for over two months in order to slowly brush her hair for her. "He is a man," this friend concluded, "who loves women." Once he built *Nautilus,* his pickup line of choice, according to another former friend, became, "This is my submarine. You want to see my submarine?" He married a filmmaker in 2011, but she kept her own residence, and they did not live together.

At *Nautilus*'s launch celebration, the spectators milled about, muttering excitedly, as Madsen straddled that submarine, once and ever after the best tool of his seduction. He kept staring—trollish, contemptuous, and eager—out into the depths, contemplating all the things suspended within the wavy dark. Inside *Nautilus,* drills screamed. Overhead, a helicopter screamed.

And the crowd gazed at the spectacle of the desire of this megalomaniac, picking his nose, gassing it up. Madsen palmed his periscope and licked his chapped lips as the Danish flag slithered. Spectators stood up in their motorboats and spread their legs apart, bracing themselves. The water gurgled like acid reflux. On the dock, beneath the hooks and the hoists, people beeped their car horns. In the water, people beeped their boat horns. It was hard to tell which was which. The wind swept it all up, allowed the noise to commingle and hover as if some cloud of pollution.

On the rear end of *Nautilus,* a wraithlike cello player bowed some horrible dirge. The song sounded like the plaintive moans of some forsaken leviathan, before accelerating into action-movie-trailer movement.

Madsen climbed back into the submarine, only his head and shoulders remaining visible to the crowd. He was eager. His hair was a mess, a boyish blond bowl cut. A barrel-chested crewmember in a black T-shirt, black cap, and sunglasses clung to the side of the vessel, grasping a bar with one hand and raising a silver whistle to his lips with the other. He had an earring in his left earlobe that looked like a washer. He shouted something—the gutturals and hisses of which carried the weight of some swaggery commemoration. He blew the whistle, and the sound cracked like the voice of a pubescent boy. Some of the other crewmembers laughed at him, gave him shit for such a weak sound. He blew the whistle again, and there it was—nice and piercing. He took it from his mouth and nodded, and the other crewmembers nodded in response. He raised his whistle hand to the crowd in a stiff-armed salute. Ropes were ceremoniously cut. The orange hatch slammed shut. Someone played a saxophone. Someone silenced their daughter, as Madsen went under.

# 8

The portion of the Øresund Strait separating Trelleborg, Sweden, from Refshaleøen, Denmark, is a phallic-shaped mini water body within the portion of the Baltic Sea that the Danes have named the Kattegat (depending on the translation, Cat's Throat or Cat's Asshole), due to its extreme narrowness and the resulting vulgarities of the captains of medieval trading fleets who, in likening their boats to their own dicks, complained of the difficulties of navigating such a tight opening. Due to these difficulties, this particular passage has seen its share of shipwrecks.

The remains of the German coaster *Johannes L* lie only three hundred meters offshore, only one meter deep. As such, the sunlight can still reach it, and the wreck appears shadowy, backlit amid the murky green water. It is beautiful and boxy and has been resting here since 1981, having gone down thirty years after its birth in postwar Germany. The merchant ship enjoyed its maiden voyage in 1951, just as Harry Truman signed a proclamation in which he officially declared an end to the United States' hostility with the Germans. It was one of the first coasters to have been built without armaments. Now, as it lies immersed in the strait, mussels have overtaken many of its masts, beams, and struts, and so at a distance the remains appear hybrid: man-made and natural. The small opening in its flank may lead to an old engine room, but also a throat, a belly, a heart, viscera. Only one meter down, one can fancy oneself a modern-day Jonah. One could lift one's head to the surface and, at will, be reborn. According to the Divescover tourist website, "It is possible [for divers] to stand on the wreckage with their heads above water."

As recently as August 3, 2018, the drunken captain of the cargo ship BBC *Lagos* found the meanders of the strait too much for his abilities, and with one hand on the wheel and the other on a bottleneck, he steered his vessel, and the five thousand tons of Lithuanian wheat in the hold, onto a sandbar over five hundred meters from shore. So beached, the captain shrugged, sipped, and sang songs to himself, and, intoxicated, ignored the calls coming in over the radio from local authorities. When the frustrated authorities finally boarded their own boat and motored out to the wreck, the captain waved his bottle, welcomed them aboard, and vomited onto their white shoes.

And the wheat roiled like smoke from a rupture in the hold into the depths, dusting the remains of the HMS *Magnet,* interred here since its wreck on January 11, 1809, a cold, cold Wednesday. The vessel had been a baby, only fourteen months old, when, overwhelmed with ice, it broke itself against a shoal and spilled its cargo into the strait— a confetti of corn kernels (the corn having been stolen from a captured Russian vessel) and a bunch of guns; linen shirts and linen pants, wool caps and jackets, silver pocket watches on serpentine chains. Still today, military history buffs dream of sinking to the seafloor and exhuming these treasures—especially the guns—but as yet, they have no submersibles of their own, and even if they did, the conditions at depth here prove volatile, and the old weapons remain buried. The four most noteworthy things to have been recovered from the wreck are three human skeletons and one remarkably and inexplicably preserved body of a sailor, his skin still luminous, his eyes soft and alert, his hair neatly combed.

Many who still live here greet the sea and the strait with an odd brew of fear and reverence, a siren from whom they cannot turn, knowing that their attraction will be their undoing. For a time, at least until she moved from Trelleborg to Refshaleøen in March 2017, Kim Wall and Peter Madsen were separated by only twenty-five kilometers of water, and the centuries-old corn and guns, clothes and watches, stories and bodies sunken into it.

≈

In March 2017, Refshaleøen was cold and rainy. Just before her birthday on March 23, Kim moved into a small second-floor apartment in a squat redbrick three-story building with her Danish designer boyfriend, Ole Stobbe. They had been together for five months. Kim was by now a renowned journalist, having written articles and essays on everything from climate change to nuclear weapons testing, from voodoo communities of Haiti to covert television delivery networks in Cuba, for outlets such as *The Guardian, The New York Times,* and *Slate.* Her journalistic eye was generous, empathetic. Of her subjects, she noted things like "his T-shirt was drenched in sweat, and he used a corner to dab his forehead." Once, in an apartment in Havana, she took the time to notice "a floral armchair was so worn and grimy that it was sticky to the touch," to notice the "tub of dulce de leche ice cream" clutched by a tired man who "fell sideways into" that floral armchair. Writing of Idi Amin's torture chambers, she traveled to Uganda and took photographs of Mengo Palace "overgrown by banana trees and papayas." There, she noticed how "the red dust swirled" around the "ballerina flats" of her interview subject; how the "burnt-out carcass of a Rolls-Royce . . . rest[ed] among grazing goats"; how, on the palace grounds, "human bones kept showing up." Once, she turned her eye to faces and to furniture, to shirts and shoes and ice cream melting. She thought of these things as important. She remembered them, and she wrote them down.

The windows of her new Refshaleøen residence were bordered by white-painted frames that, from a distance, looked like bars. If it weren't for the overcrowded bicycle rack in the front, the place—and indeed the entire neighborhood—could have been mistaken for a prison yard. Kim's building abutted a defunct gray warehouse, the facade of which was discolored with stripes of mold, and smoke-stained. More bicycles had fallen to the asphalt than remained upright, still tethered to the rack by their locks. Kim's apartment window was located above this rack, and if she were inside working on an article, she would have often heard the metal-on-metal scrape of bicycle toppling against bicycle, the other residents cursing as they worked to unearth their rides from the bottom of the pile, wrestling handlebars from chains, pedals from spokes. She would have listened to this as her hands hovered over the keys of her laptop and her fingers cast long shadows on the wall, as she wrote of ice cream and bones, of human beings and the things they loved and endured.

One night after dinner, she and Ole strolled among the alleyways of the island. When they passed the amateur rocket-building workshops of Copenhagen Suborbitals, Kim's curiosity sparked. She took cursory notes and, over the next days, made a few inquiries. She had traveled so far to uncover her stories, but now there was a story hatching in her backyard. She soon heard of Peter Madsen. Rumor had it he not only built his own rockets but was into homemade submarines as well. Kim tracked down his contact information and left numerous messages, but Madsen did not return her calls or texts. Kim put the story on the back burner, but continued to pitch her idea to magazines and newspapers, hoping to profile the neighborhood eccentrics and write on the "space race" between Madsen and the Suborbitals.

Five months after moving to Refshaleøen, exhibiting her journalistic spirit of being drawn to "the undercurrents of rebellion" (as she defined it), Kim and Ole decided on a whim to move to Beijing, where he would attend university and she would hustle for freelance gigs. Online, they rented a house there, sight unseen. Kim began to conceive a long-form article on the neo-Maoist movement and started searching for Beijing-based interview subjects. On August 10, a Thursday, Kim and Ole hosted their own farewell party for a small group of friends. Together, they talked and laughed and worried, ate hors d'oeuvres and drank and danced in the soft yellow light of the end-table lamps.

A few kilometers away, the first act of Copenhagen's Egmont Festival—a three-day outdoor musical extravaganza for an audience of predominantly university students—took the stage. It was the local sensation, Marvelous Mosell, a Danish version of Vanilla Ice, who began rapping poorly and dancing poorly in his baggy purple windbreaker, baggy purple parachute pants, pink sweatband, and yellow wraparound sunglasses, carrying around a boom box of exaggerated size on his left shoulder, his chunky chain necklace swaying (bearing a fist-size MOSELL pendant rendered in bubble graffiti lettering), his giant pinky rings reflecting the fuchsia stage lights, casting a glow onto the beatific faces of the crowd. Mosell had recently updated his Facebook page with the directive "Check out the rhymes I bust!" and the crowd of college kids—immersed (according to the Egmont Festival's own press release describing the atmosphere of the event) in an "absolute daydream climate of sweaty bodies, cold beer, and wet kisses"—complied.

Through this crowd, Peter Madsen, on his way from the seaside to his workshop, clad in a jumpsuit the color of beach sand, snaked. Based on his former friends' assessments of his personality, Madsen likely felt contemptuous of the concertgoers, but kept his feelings of superiority to himself, his elbows pasted to his ribs. Though he could have used a more roundabout route to his workshop, avoiding the festival crowd, he felt compelled to take the shortcut.

Onstage, Marvelous Mosell performed some clumsy break-dance maneuvers. Weaving through the crowd, Madsen clutched his cell phone, scrolling through the pages of his crowdfunding accounts, which in the preceding days had proved at first to be a source of anxiety, then anger. His plan had been to stage the big reveal and subsequent public launch of his latest in-progress rocket (which he christened the Alpha, formerly "Imperial Star Destroyer"), an event that would surely cement his technological and intellectual dominance over those philistines at Copenhagen Suborbitals, and represent the nail in the coffin of that company, and the public shaming of the former colleagues who had forsaken him. He had scheduled the event for August 26—sixteen days from now. But the funding he was anticipating had not rolled in this time, and he tried to prolong admitting to himself that he was going to have to abort the launch. He was crestfallen and frenzied. Despairing and irate.

He was desperate to prove himself, test himself, see if he was still capable of pulling off the extraordinary. Madsen took to scrolling through his older text messages, on the hunt for missives praising his ingenuity, testifying to his importance, believing him to be worthy of attention.

Amid the herky-jerky movements of Marvelous Mosell's youthful audience, the glow of his cell phone lighting up half of his face, Madsen searched one crowdfunding account, then another, hoping for some last-minute surprise windfall, some anonymous donor, seduced by his clear genius, who had decided to finance his launch, and his "victory"—finally!—over the Suborbitals. But, no. Nothing.

The amounts remained unchanged and insufficient. If he was honest with himself, he was expecting this. He had already decided he would need to otherwise channel his growing rage. He would have to otherwise restore his dominance, his superiority. He bit his lip and allowed his elbows to rise, shift the sweaty bodies around him. He closed his crowdfunding pages and reopened his text messages. He flicked his callused thumb upward along his phone, and the older messages rose to the top of the screen, until he located the one he wanted—the one from the local journalist. He jabbed his thumb into the screen's center, and the upward scrolling stopped. He wore a faded green backpack, the top of which was unzipped to accommodate the length of the orange-handled handsaw he'd stashed inside, and one concertgoer, whose sweaty torso was the recipient of one of Madsen's errant elbows, would later report having seen the rusty silver blade of it, the jagged sawteeth.

Back at the farewell party, Ole was probably flipping burgers and toasting buns on the coals as Kim mingled, replenished her guests' glasses. No doubt they could all hear the dull thrum of the music in the distance, the wind carrying the vibrations of the Egmont Festival along the water. At 6:45 p.m., Kim's cell phone vibrated. To her surprise, it was Peter Madsen, inviting her to tea at the nearby hangar where he lived and built his machines. When she texted back, asking about a good day and time, he replied that right now would be perfect, but just for tea, he stressed, just for a cursory conversation about setting up a time for the full-on interview. She checked in with Ole and told him she would rejoin the party later. It was a short stroll—only a few hundred meters—to Madsen's workshop, and Kim, in her excitement, walked quickly in the waning light, the orange glow of the streetlamps reflecting from the corrugated iron facades of the warehouses, the piles of old ship parts. Her hair was tied up in a bun. So many Danish newspapers later seemed eager to report on this detail, and when the articles were translated into English, the hair bun came up as "potato." *A potato on her head,* they said. *Potato, potato, potato.*

# 9

THE COOL COPENHAGEN EVENING wind animated the strands of auburn hair that had loosed from what the Danish-newspapers-in-translation would later call Kim Wall's potato, and this is important: The way in which wind can inspire our hair to whirl about our faces is important. The way in which parts of our bodies can be tethered to us is important. The way in which we are compelled to raise our hands to our faces, to sweep our hair from our eyes when the wind forces it back down, is important. Our hands lifting from our hips, rising to our faces, seems a small gesture, but it is not. We agitate the atoms, the components of the air we breathe, the insects too small to be seen; an entire cross section of a world invisible to us is disturbed. Electrons tumble through space, gain their bearings, and rush to rejoin their clan. And they do this because we move, and we move because we are alive.

Kim Wall moved. And the wind moved her, as she made her way to Madsen's workshop. The smokestacks, power grids, shipping containers, and gray concrete silos turned the cracked streets into a labyrinth. The dumpsters were painted bright green. The graffiti was cute and benign, childlike: rudimentary owls and bats with goofy grins, hearts and initials; rainbows and stick figures and rockets and boats. Those bone-like windmills spun in the distance. Chunks of concrete and cinderblock and rebar decorated empty parking lots. One knoll sported wildflowers, another a graveyard for construction cones.

She would have wandered past the crumbling redbrick facade of the Asterions Hus theater, the street in front of it harboring puddles, even when it hasn't rained. Inside, no doubt, the tireless troupe was rehearsing their performance, *Meeting the Odyssey*, which fused elements of Homer's classic with "contemporary themes relevant to the people living in Europe." The performance was to take place on a "sailing ship" and involved music and dancing and poetry and the releasing of balloons into the air.

Maybe Kim was rehearsing the questions she intended to ask Peter Madsen. But maybe her mind was elsewhere, beautifully drifting. Maybe, as we are prone to do when wandering the streets alone, she took stock of her life, all the birthday parties thrown in her honor. All the balloons and candles blown out, and faces cracked with laughter, and her name, scribbled in so many colors of icing over the tops of so many cakes. Maybe she thought about tigers, which she had loved for as long as she could remember. Their eyes, their paws, their whiskers, their stripes. Maybe she recalled that moment when, on assignment in a remote mountain village in North Korea, a group of women surrounded her and couldn't keep themselves from playing with her hair.

It was getting darker, and the world was bluing. She walked past the marine supply shop with its broken windows, the frames of which were sealed with Visqueen or foil. She walked past the overgrown lot that would serve as the site of the Copenhagen Photo Festival, where artists would mount their work on the flanks of the shipping containers and erect huge easels and displays amid the tallgrass and the weeds, the crickets complaining. There, on that patch of flattened thistles, in only a handful of months, would be a giant, controversial photo of a woman's face—eyes closed, mouth open—cinched into what appeared to be a plastic bag.

How unencumbered she must have felt on this assignment to interview Madsen—so close to home—compared to her overseas assignments. She often brought extra luggage so she could bring gifts home to her family—elf figurines and bright red socks. For her parents once, an encyclopedia on South African history. For her brother, Tom, a home beer brewing kit. Kim's mother once joked that her suitcases weighed more than Kim herself.

Kim walked past Alchemist, the fine-dining restaurant housed in what looks like a former rendering plant. She hung a right and walked past the network of outdoor courts where families and coworkers played sunset paddleball, sweating and letting their sweat cool. She walked past the former B&W Hallerne industrial complex, which looks like the biggest headstone in the world, and which, three years prior, served as the host site for the Eurovision Song Contest.

She passed the industrial-chic modern art venue, and the concrete monstrosity now called the Grand Ballroom, and the Paintball Arena. Kim took a left on a street with no name—a sorry excuse for a street, the asphalt sandy and soggy, the roadside resembling some mad scientist's junkyard, littered with rusted beams and hooks, springs and drums. Up this ruined road, Kim walked 160 meters. Her shoes hovered over the puddles. They left their tread in the sand.

Madsen's lab was housed in an annex of the old Horizontal Assembly Building in the defunct Burmeister & Wain shipyard and was cobbled together from three different colors of corrugated iron. The upper half olive green, the lower half baby blue, and the door, a darker blue— the blue of the strait. On the door was a crooked white placard bearing the address: 185. To the right of the door, nailed to the baby-blue iron was the sign RAKETMADSEN'S RUMLABORATORIUM—Rocket Madsen's Laboratory Room. The iron around the door was dented and pocked, scarred with wayward nails and industrial staples. Pillars of black mold climbed up from the bottom edge. Dandelions lived and died at the base of the lab, the border where the iron met the loose asphalt. To the left of the door was the faded blush of what appeared to be red spray paint, and to the left of that was a curvaceous white mailbox, the front slot held on with two different kinds of tape. Pasted over a strip of packaging tape was the circular blue-and-yellow logo of RML Spacelab (Rocket Madsen Spacelab), the company Peter Madsen forged after his fraught departure from Copenhagen Suborbitals. Shooting upward from the trench between the two towers of the M was a conflagration meant to represent the plume that blossoms from the ass end of a rocket during liftoff. Above the mailbox was a yellow rectangle, printed with the warning *Trykflasker fjernes ved brand*—

"Pressure bottles are removed in case of fire"—beneath the image of an exploding canister and a furry-looking flame.

Kim stepped to the door.

≈

In a matter of months, after the horrid details of the crime against her had come to light, Refshaleøen—all that she had just walked through—would seem to be saddled with the musty mournfulness of a wake. When I walk these streets and alleys and open lots six years later, the place still bears such a hush. It feels played out—not sleepy in some quaint way; just tired. Not only had the old industry exhausted itself, but it seems as if the once-renegade drive to repurpose it has as well. More two-story steel shipping container and corrugated studio apartment complexes have sprouted up like hipster prisons in previously vacant lots. They are sad rectangles amid the broken glass and the soggy gravel, the barbed wire and cumulous tangles of flowering weeds. They are fronted by so many toppled bicycles, and no people.

A rotting bus bridges two green shipping containers, evoking a junk archway that, passing beneath it, heralds another open lot, kudzu sprouting from the gravel, leading eventually to the quay, where the intrepid bungee jump from a giant old crane over the leaden arm of the Øresund Strait, the four distant smokestacks belching uniformly, the smoke arcing like four perfectly groomed eyebrows. I feel surveilled, as if by the security force of some corrupt company town that has secrets to keep. When the wind dies down, the place is so quiet that the rare car or bicycle tire humming along the rain-slick streets is loud enough to unnerve. Even the few pigeons won't let me get within fifteen feet of them without bursting into the air, flapping toward the canal, which itself is the color of pigeon feathers.

Refshaleøen's residents seem to receive their mail on one corner—a cornucopia of metal P.O. boxes mounted on the side of a brick wall and an attached sheet of plywood. One person's address, scribbled in gray grease marker, simply reads, *This Guy!* In front of this ersatz post office, an orange-and-black sign welcomes one to Refshaleøen with three strict rules: (1) Ban on electrically amplified music from 10 p.m.

to 8 a.m. (2) Fires are forbidden. (3) Show consideration and restraint. Next to this last item, someone has pasted a sticker reading, A CROWS MURDER. On a nearby shipping container, a wheelless, propellerless helicopter petrifies as if the head of some giant shrimp-god.

*Oen*—"the island"—it reads in yellow cursive paint on the side of Copenhagen Suborbitals' missile-like rocket HEAT-2X, decommissioned and mounted in the lot outside its green-and-black iron warehouses. It is about twenty-five feet long and skinny, and its tip is orange. The rain drips from it in thick strings—almost like a gel—and the thing is filthy, spotted with a mucusy moss. The earth here is repurposing that which we've repurposed, the concrete and iron by-products of our industry and our art. Refshaleøen is how I'd imagine our neighborhoods to look fifteen years after the extinction of the human race. Our dumpsters will survive us. They will be our evidence and testimony. Here, they already are.

Across the flooded gravel lot from Copenhagen Suborbitals is Madsen's old laboratory, where Kim Wall stood six years ago, about to knock. Now the lab seems like a sad and slapdash mother-in-law studio erected in the Suborbitals' backyard, as far as possible from the main house. Vandalized shipping containers rise from behind the place, blue stacked on white stacked on blue, painted with fists and fingers, pentagrams and halved eggs. On an adjacent electrical box, someone has spray-painted FUCK NORMALiTiES! This seems like something Madsen, or one of his followers, could have written. The laboratory faces the rear of the B&W Hallerne, the giant concrete ex–shipbuilding warehouse on which a two-thousand-square-meter black-and-white mural of a wolf's eyes and muzzle was painted in 2015 in honor of the Copenhell festival. Some insist that the mural is called *The Wolf of Copenhagen;* others say it's called *Hunted.* But most refer to it as *Fenrisulven,* or "Fenrir's Wolf," after the wolf in Norse mythology that bit off the hands of the fearful gods who tried to bind it. The wolf here still serves as a symbol of wild ferocity, the wolf that will grow large and powerful enough to break free from its constraints and ultimately kill Odin, the one-eyed omniscient All-Father. This is the two-thousand-square-meter wolf Madsen stared down each time he left his workshop. This is the wolf that still looms even now, six years after Wall's death. *Heed,* it says on the workshop door in drippy red paint. *Help,* it says.

~

I catch my breath on the side of the B&W Hallerne warehouse, around the corner from the wolf, at the feet of a spray-painted Yoda in a red smoking jacket, lording over the caption SYSTEM-BEVARENDE MID-DELKLASSE (*System-Preserving Middle Class*). The asphalt here, wet. The pompons of weeds, wet; their hairlike thorns glistening. There are more mosquitoes here, it seems, than anywhere else in the city. Old, unused railways are embedded into the streets.

I emerge from the side of the warehouse. I take inventory one last time before turning my back on Madsen's old workshop.

In a neighboring alley, a six-foot-tall wooden crucifix balances against the seat of a parked baby-blue bicycle. Next to it, a man squats in a steel-gray jumpsuit, snaking his fingers through the links of a door in a tall metal fence, the top half of his ass cresting his waistband. It looks as if he's trying to break in. On the other side is only another sodden, weed-pocked lot. He's ignoring the red-and-white sign zip-tied to the fence, an image of a person in shadow, raising a giant hand to stop whatever and whoever may be approaching. KUN ADGANG FOR KORERSKOLE ELEVER, the sign warns—*Access Only for Choral School Students.* SET-UP, it reads in huge spray-painted bubble letters on the outbuilding on the far side of the lot, behind yet another fence topped with barbed wire. Old, defaced signage reveals that this space may once have been an experimental film studio, adjacent to Copenhagen Suborbitals. But now it appears to be a part of the Suborbitals' property.

Six years ago, in 2017, Kim Wall stood at the door to Madsen's lab. The gravel crunched under the toes of her shoes. She had her questions ready. Kim knocked, and her knock sounded like any knock against a door of corrugated iron—hollow and resonant and distant.

# 10

My attempts to track down the PeaceSubs consortium run into dead end after dead end, until I realize I misunderstood Hank Pronk. The organization is actually PSUBS—P, not Peace—or the Personal Submersibles Organization of Amateur Submarine Builders and Underwater Explorers, and they hold yearly conventions.

When thirty-some personal submersible enthusiasts—all white men from their twenties to their seventies—converge in one place, and when that place is an ill-kempt Hampton Inn in the middle of Michigan, hijinks ensue.

It is a wild mélange of T-shirts and polos and sandals and cargo shorts. A little denim, a lot of khaki, a preponderance of beige ballcaps with salt-stiffened brims. Their waistbands are stained with old sweat and their shirts crusted with old deodorant beneath the short sleeves. Many of their shirts have the word *Crew* on them. On their breast pockets are anchors and American flags, sheriff badges and life preservers, cursive designations like *Chief* or *Captain* or *Safety Officer*. They wear their clothes one size too big. There are lots and lots of blue eyes. These men wear wraparound sunglasses on their faces or on the backs of their sunburnt necks. Some wear those water socks with individual toes. They shuffle from foot to foot in the parking lot, and they squat on their flatbeds and they emerge like spring-loaded snakes from their submersible hatches. Many of their trucks are equipped with CBs, walkie-talkies. They make faces at one another through the convex panes of their portholes. They are experts at driving in reverse.

They are dentists and engineers, mechanics and "self-employed";

fathers and grandfathers and confirmed bachelors, divorced, divorced, and divorced. They are self-described "misfits." They live on and off the grid. They have a distrust of government ranging from healthy to conspiratorial. They cruise for parts on Craigslist and in junkyards. They overuse words like *cool* and *ultimate*. Some have been working on their subs for years but have never actually dove. Some spend more time underwater than "up top," as they say. Some of them never don't have a can of beer in their hands. Some compliment one another on new beards, moustaches, and sunglasses, and some rib one another for the new gray, the new crow's-feet, the baldness that has expanded since last year's convention. They commune about their drives in, the subs they towed that made them spectacles on the road—other cars honking and waving. So many thumbs-up, so many *hang loose* shaka signs. "You pull into a gas station, and it's twenty questions," one says.

But these pleasantries are short-lived, and the conversations quickly take a turn toward the geeky and the niche—types of welding torches and frames and epoxy and acrylic. They pass around tools and air tanks and Body Glove paraphernalia, and show one another cell phone videos of themselves welding something or tightening something or loosening something. They are never not rapt with these videos.

Some show videos of their previous dives. Some have scored their videos with AC/DC songs, some with Enya. Some feel compelled to narrate: "We are in the sub. Personal sub. This is a movie from inside the submarine of the fish that are swimming around." Some feel compelled to sign off, "Over and out." In most, though, all you hear is the wind. They watch and heckle each other's mistakes. "I don't see no bubbles! I don't see no bubbles!" they jest when a sub struggles to sink below the surface. "How's *that* for buoyance control?" someone else boasts. "Shoulda started the clock so I knew how much air I had," another self-critiques. They perv over things like visibility and control and stern thrusters. On another undersea video, a p-subber refers to a starfish as "that little bastard." "But its genus is *Pycnopodia*," he follows, prompting the men to debate the difference between starfish and sea stars, the actual number of species of the fish we blanketly call red snapper. They call out the images of garbage on the seabed. "There's a bottle," says one. "There's a tire." Many refer to sub parts as "sexy."

The steering column—sexy. The rudder controls—sexy. The curve of the acrylic window, which has that added juice of danger as, under pressure, such windows don't gradually crack but instantly implode—sexy. They make obscene jokes about the size of a sea cucumber, "wet dream" puns.

Most of the videos are terribly boring. In some, babies cry in the background off camera. In most, "uh-oh, uh-oh" is a common refrain. They all call oxygen *Oh-2*. They say *foot* when they mean *feet*. "Damn," the men say as they watch.

The Hampton Inn parking lot is littered with oversize vehicles and trailers, and these little personal subs. Subs, sprouting from the asphalt like tumors. Most are yellow, but some are blue, or white, or red and white, or red, white, and blue. Their captains pose for photos in front of them, affecting two and only two postures: the stoic straight-backed stance, hands ceremoniously crossed over their pubes, and the squatting-to-the-asphalt-on-one-knee pose, elbow resting on the lifted knee and the beer can in that hand, the other hand resting coquettishly high on the thigh of the asphalt knee. Those who affect the latter pose will have red indents on that knee through lunchtime. Sometimes, the men remove their ballcaps, as if before a queen, and sometimes the ballcaps remain on their heads, shielding their baldness from sunburn. Their subs are named *Snoopy*, and their subs are named *Challenger*, and their subs are named *Special K*. These men cherish their cartoon dogs and their rocket ships and their breakfast cereal. One of the men has named his sub—one of the smallest in the world—*BIG*. *BIG*'s propellers, someone says, are so tiny they probably get caught in pondweed. One aerodynamic-looking silver sub named *Git Kraken* resembles the beaky head of a *Spy vs. Spy* protagonist. It stands out among the more conventional designs of *Persistence* and *Sgt. Peppers* and *Nemo* and *Bionic Guppy*. They photograph their subs and will later upload these pics to the PSUBS webpage.

I'm not sure how often the webpage is updated, but still there, haunting the site as of 2023, are images and videos of Peter Madsen's old sub *Kraka*, and his *UC3 Nautilus*, diving off the coast of Refshaleøen. A couple of the videos are preceded by ads for a "real-life light-saber"—a military-grade flashlight called Elite Tac, designed for

American Special Forces "to view enemy positions from miles away." ("It couldn't be sold to civilians... until now! The government is trying to classify it as military-grade only, and it probably won't be long before it's taken off the online market. But be warned! This is not a toy! It is so powerful, it can melt through anything! It literally burns any material in seconds! It can even start a fire! It is completely indestructible! It is the perfect tool for survival and self-defense! It will even scramble an egg!")

In these videos, in such different times, families stand on the dock with blond children on their shoulders, taking pictures of Madsen's bygone descents and clapping their hands as the water bubbles up along the sub's gray flank. In some videos, Madsen emerges from the hatch upon surfacing in steely military fatigues and cap, rests his elbows on the edge, and stares at his hands, ignoring the onlookers as they applaud. In others, he ignores them while doing calisthenics on the hull—jogging in place, stretching out his arms. "Peter! Peter!" they call to him, but he pays them no mind as he haunts the PSUBS site, occasionally bringing his fingers to his lips as if he's smoking some invisible cigarette. Most everything in these videos is the color of pencil lead. Sailboats loom behind *Nautilus* like dorsal fins, and behind them, the buildings of Copenhagen rise, dark and industrial, as if some ash drawing of a city.

Here, in the Hampton Inn parking lot, the men use what idle time they have to catch up, to shotgun their beers, to brag about their mechanical prowess. They take new pictures and new videos. They make new memories to redact the old ones. They pack their days with distractions. The three-day convention event schedule is vague and succinct. The 8:00 a.m. Lobby Meeting is followed by two hours of Travel (to a lake for their dives), which is followed by Lunch. At the PSUBS conventions, the restaurants of choice are pubs called the Rusty Gull, diners called the White Spot. Jokes about the menus' lack of submarine sandwiches abound. The veterans of the group groan the loudest at these jokes. They've heard them all before. The rest of Day One is comprised of a Business Meeting (Members Only) and a Rum Tasting in the hotel lounge. In the bars and restaurants, they forsake the tables for the booths. They take all the booths. They race for them.

They fight for them. These are booth guys. On the final night, they will eat burgers by candlelight.

At these conventions, the featured speakers' topics range from underwater innovation to nuclear technology on old sunken submarines; from "rescue and/or escape from personal submersibles" to "Syntactic Foam: What it is and how do you make it" (first, you need to get your hands on some microballoons . . .). The men will gather in a spartan ballroom to watch rickety film strips from the 1960s and '70s featuring an early hero of the personal submersible world named Mart Toggweiler, a German enthusiast who moved to Long Beach, California, and initially gained attention in the community for his seventy-two-page booklet "How to Build an Underwater Camera Housing." After the booklet's publication, Toggweiler, with his toy sub, *Submaray,* landed contract work from various governmental organizations— inspecting subsea power cables, recovering or confirming the location of the remains of missing persons, searching for bombs accidentally dropped into the ocean by military aircraft. According to a journalist covering Toggweiler's missions at the time, *Submaray* was so small and flimsy-looking that "conceivably a large octopus could tote it home for the kids to play with." Another said of the sub, "Conceived in a backyard project, [it is] undoubtedly the ugly duckling in a pioneering field."

Nevertheless, *Submaray* persisted. Using his underwater camera housing, Toggweiler documented his misadventures in the sort of murky visuals and monotone voice-overs that make one not quite nostalgic for the sex education film strips of the 1950s. Onscreen, when *Submaray* encounters a dead whale floating on the surface, picked at by gulls and sharks, the narrator flatly intones, "The odor was intense, but by approaching upwind, we got close enough. It was difficult to realize that this bloated heap of rotting flesh had once been a huge, graceful mammal."

The men featured in these documentary shorts have close-cropped haircuts and go on their diving missions wearing horn-rimmed glasses, snug white slacks, and cardigan sweaters. They communicate with one another via two-way radios with four-foot-long antennae. The dock crew are dressed like hippie cowboys, in leather boots, wide-brimmed hats, and tight denim ranchers' shirts with the top three buttons undone. They communicate via chunky black telephone receivers, cigarettes dangling from their lips. Proto–baby boomers, all. The viewing audience in the ballroom murmur and giggle. Afterward, they race for the doors, congregate in the parking lot to puff from their Camels or Marlboros, or stubby ends of fat cigars left over from earlier. In talking about the film, some pine for the "good old days," and some make snarky jokes. No matter their reaction, when the film strips end and the screen goes blank and bright white and stroboscopic, the men seem ready to go on their own misadventures.

At a previous convention, the undersea technology company Nuytco provided convention-goers with two well-appointed submersibles—the one-person *Deep Worker 2000* and the three-person *Aquarius*—for the attendees' diving adventures. Many of the PSUBS members' submersibles—however beautifully they ornament one hotel parking lot or another—are in "development" stages and unsafe for the water. Nuytco was perhaps most famed for its fetishistic Exosuit (the newest generation of its earlier Newtsuit), a metal diving suit that's part superhero getup, part edema mannequin, part Michelin Man.

The men posed with *Deep Worker* and *Aquarius,* sure, but most were excited to have their picture taken with Exosuit, so they could later post their photos to their social media pages with captions like *Up close and personal with Exosuit, Holding hands with Exosuit,* and *Unfortunately, I think Exosuit is already spoken for.* One plus-size bearded gentleman in a sky-blue T-shirt and beige ballcap posed with the suit and later posted the image, lamenting, *Hmmmm . . . We're either going to need a wider Exosuit or lots of butter to squeeze me inside this one. Hey Phil, got something in a 44 extra-extra-large??* Phil was Phil Nuytten, owner of Nuytco—a billionaire sponsor, diver, inventor, and self-proclaimed "submarine guru" whose submersible designs had been solicited by over a dozen worldwide navies and who, when asked why, as such a professional, he continued to encourage this particular group of amateurs, answered, "Because I have known what it is to be thirsty, I would dig my well where others may drink."

For Nuytten and his PSUBS disciples, the sea (or for some, the lake or the pond) represented the sort of unexplored frontier that can be difficult to find on land, unless one wants to disappear into the mountains or swamplands, set up shop on the wind-battered tundra, or hang out on a glacier. In a p-sub, one can, unregulated by any governing body, surf the unknown while simultaneously armored against the unknown's discomforts. How strangely American this seems, this pursuit of some hackneyed idea of liberty and pioneering from the vantage of a sort of self-designed La-Z-Boy. Navigation by remote control. A beer within reach. Fascinating how the remote control has impacted our fresh conceptions of manifest destiny, as we dream of building our cloisters on the moon, or in outer space, or under the sea, pouring our water-safe concrete into the pores of the coral reef in order to actualize them.

For the p-subbers, the naval submariners, the ultra-rich contingency, and the wider world of enthusiasts, the sea is a fail-safe—a place to which those prescient enough to have the machines to pilot themselves there can retreat and start fresh once we've sufficiently depleted the land, once the climate, with our participation, conspires to make the surface less comfortable for us, and so many others. In this way, we can flee that which we've irreparably wrecked, then begin our slow wreckage anew. Like Columbus. Like John Smith. This, the p-subbers may tell themselves, is what our founding fathers would have wanted, the westward expansion moved downward. A fluid, gridless place to test their spirit and their backbone, a place in which to disappear, a space on which, at long last, they can't possibly be trodden.

Nuytten embodied this ideal, and his innovation and wealth inspired the other men to believe he could make it happen. They treated him as a wagon-train leader and guru, and he took naturally to both roles. Nuytten had the fixed gaze of some dyspeptic hunter and the salt-and-pepper hairdo of a TV news weatherman after having reported in a hurricane. It appeared windswept and flattened, but also as if it had once been intricately sculpted and larded with plenty of product. Sometimes he'd point his hairdo toward the nearest body of water, train his gaze on the horizon out to sea, and, conjuring his inner soothsayer, utter in a faraway voice such maxims as "We should be there, and diving..."

Attendees rhapsodize about *possibilities,* some in reference to a recent announcement by NASA that it would be funding the construction of a submarine to be blasted into outer space to explore Titan—Saturn's largest moon—the sole place in our solar system, besides Earth, that has standing bodies of river-, lake-, and sea-like liquid on its surface. It's also believed to contain a subsurface ocean. NASA describes Titan's atmosphere like a paint swatch—"Golden Hazy." The proposed space sub (tentatively christened *Titan Sub*)—cylindrical, and about the length of a station wagon—would drop through this golden haze and dive into Titan's largest liquid hydrocarbon sea, Kraken Mare, where it would meander for ninety days in the −300°F brine, creeping along at a meter per second until it covers two thousand kilometers, all the while sending images back to Earth on its search, in part, for signs of extraterrestrial life.

Apparently the sea is prone to some wicked whirlpools, has an area of constriction and intense currents we've named the Throat of Kraken, and is home to unexplained phenomena of "changing features" that NASA calls the Magic Islands. According to NASA's computerized simulations, *Titan Sub*'s control panel will sound like R2-D2, and its "eye" will process and analyze stimuli like that of the Terminator. The proposed landing (or "splashdown") date is 2040, but one impatient team at NASA came up with the idea of reflecting sunlight onto portions of Titan in order to warm and light up the cryogen, make exploration easier, and move up the date, but that, of course, could prove catastrophic. The potential for a new "space race" is in the air, and NASA wants to win it for "our nation." The PSUBS guys are aggressively excited that it may be a submarine—a rocket-submarine; an *American* rocket-submarine—that will finally uncover evidence of aliens.

In the face of such magnanimous unknowns, one may seek refuge in the known, or at least the educated guess. Due to its preponderance of methane—and akin, no doubt, to many of the personal submersibles here—Titan smells like farts.

In the trees at the edge of the parking lot, blue jays are fighting. A woodpecker feeds, keeping time. There's more bird shit on the subs than there was ten minutes ago. The pleasant release of beer cans being cracked, one after the other, offers a sonic alternative to the woodpecker's testimony. Most of these men seem to have excellent self-images. Via the avatars of their subs, and the personalities they heap upon them, they see themselves as extraordinary, resolute enough to stand outside the bounds governing ordinary society. "The common person does not have a submarine!" one of the men shouts. Yep, this is quite the happening, and the more it happens, the more one may become curious as to how it happened in the first place.

≈

Jon Wallace was a bored computer programmer for a New Hampshire branch of Hewlett-Packard who, after spending thirteen years of his free time in positions of local government (serving on the Community Access Television Committee among others), decided to foster a community around his weekend passion. Wallace founded PSUBS in 1996, building the group around and out from an initial online private chat room of only four dedicated backyard sub builders.

"When those discussions became worthy of capture and archiving," Wallace says, he decided to create the PSUBS website to track the contact information of all participants, the number of which grew to "hundreds" over the next twenty years. "PSUBS began offering formal memberships in order to provide a sense of ownership," Wallace says, one of the stipulations for membership being that the sub in question must be privately owned by an individual and "housed in their own garage." The organization started holding its annual conventions in 2002, when the predominantly middle-aged, upper-middle-class members expressed their need for it. Wallace says that these are guys "with their mortgage and kids under control," and who have the drive to leave their families and homes behind for a few days to geek out in some hotel parking lot across the country with their fellow enthusiasts. They need support. They need affirmation. They're tired of fighting with their wives, Wallace says: "It's not that easy to say, 'Honey, I just need $25,000 and the driveway for the next two years.'"

Wallace feels that it's precisely the niche quality of the obsession that makes for such a tight-knit group, and the near desperation to be around other like-minded middle-aged men with their mortgages and kids under control. He also feels that their shared braving of the dangers of diving to depth in personal submersibles-in-progress, in trial-and-error fashion, provides a shortcut to intimacy. The danger factor is why it's so important, Wallace feels, to perpetuate this community, to share knowledge of what works and what doesn't. "If you don't do it right," he says, "you're not going to come back up ... Anyone can build a sub to go down in, but if you want to come back up alive, you have to put some thought into it."

Like many of his cohort, Wallace locates the origins of his obsession in childhood, "from TV shows such as *Sea Hunt* and *The Undersea World of Jacques Cousteau*." Those early TV shows inspired him to pursue "the feeling of being surrounded by water, seeing how far I could swim underwater while holding my breath, and wondering what's down there on the bottom." He started scuba diving in the cold water off New England, but the frigid temperatures prevented him from being down longer than twenty minutes. Frustrated, Wallace began thinking about alternative ways to spend longer periods at depth. "So the idea flippantly popped into my head that I need something like a submarine that can insulate me from the water and also get me into deeper water," he says. "I mentioned this to a coworker when I returned to work who, instead of laughing, said, 'I know someone that wants to build a submarine.' That was my 'eureka' moment. I realized someone else had the same idea I did so it couldn't be such an outlandish idea after all."

Today, Wallace has amassed a PSUBS advisory board of welding enthusiasts who contract out their personal subs for use on maintaining deepwater oil fields, and storage facilities for commercial nuclear power plants, fossil fuel plants, and oil refineries, though the majority of those industries have switched to robot labor. The PSUBS group's Facebook page has a strict set of rules:

1. Stay on Topic: This group is for underwater vehicles, specifically human-occupied ones. This is not a group for Scuba diving or underwater robotics.

2. No promotions or spam: Please do not attempt to sell submersible well pumps. This is a group about underwater exploration, not farming. Personal sales of relevant objects, or corporate advertisement of relevant products is allowed. Please no scuba gear for sale.
3. Be kind and courteous: Please don't make personal attacks on other member's [*sic*] experience/ideas/beliefs/etc. Political content unrelated to the ocean, religious, sexual, or other divisive content will be removed.

Many of the group's conversations center around safety issues, and many of these posts contradict one another. A new member of the community may have a tough time discerning which advice to heed. The bottom line is, according to one member, "It's very dangerous to build your own sub. You never know what is going to happen." Tales such as the 1990 fatal p-sub accident of Interlochen, Michigan (wherein a poorly installed Plexiglas porthole caused the sub to implode while submerged in a local lake), still occasionally haunt the site. One p-subber—a dentist in his surface life (one of the professions with the highest rate of self-harm)—admits that when he got serious enough about his obsession to actually start building, "[Everybody] thought I was on the road to suicide." When another member's sub broke down to the point of irreparability, he, glorifying the Germans' ceremonies in defeat, scuttled it in the same way that "the U-boat commanders did with their subs at the end of World War II. I don't want to sound dramatic, but I was standing on the *URV-1* as it went under in a lake on its final dive." But others—the self-styled "old hats"—empower themselves via understatement, wizened beyond excitability. "It's like being in your car and looking around and seeing things," one says. "You'll see fish. You'll see tin cans."

The veteran members often dispense advice to the newbies, stressing that, when underwater, we perceive time differently, and suggesting they set an oven timer so they don't run out of air. The guys hotly debate who should be inducted into their newly created PSUBS Hall of Fame. Current inductees include William Kittredge—the late naval officer whom the group lauds for playing polo in India with the occupying royalty, and attending Gandhi's funeral; for using his submarine to expose Israel's nuclear power plant "before the CIA knew it was there"; and, upon his retirement in 1962, for becoming the godfather of sorts of personal submersibles, welding them together in his backyard in Maine—and Phil Nuytten.

*"If we can see further, it is because we stand upon the shoulders of Giants,"* proclaims the online gateway to the PSUBS Hall of Fame. "This is the highest honor PSUBS bestows and the hardest to earn. The PSUBS Hall of Fame is dedicated to those men and women who have significantly contributed to the advancement of submersible design, construction, and/or operation." The only inductees thus far are white men, and Nuytten is the only one who's still alive. (Note: Since this writing, Nuytten too has passed.)

Most of the group seem happy to have found a community in which they don't constantly have to justify their obsessions, don't constantly have to stress, one p-subber says, "No, I'm not that crazy. No, I don't have a death wish." If Wallace is proud of the community he's helped to foster, he wears his pride subtly. Unlike most of his peers at the convention, who—if they're not tumbling down the rabbit hole of some byzantine manifesto comprised of hardly scrutable tech jargon—speak more in whoops and overblown jeers and high fives and back slaps, Wallace is restrained, his jaw set, his shoulders squared, ready to deflect. Maybe he's used to it, as the spokesperson for a sub community that's often dismissed by the masses as a bunch of crazies. Even the so-called legitimate submarine community marginalizes these guys.

And so—especially in reading their contributions over the years to the PSUBS online chats—their default can be fiery defensiveness, and the doubling down on one's own ideals that can lead to a cultish breed of extremism. But Wallace is careful to play any extremism he may possess quietly to an outsider. He wants to offer no one an excuse to think less of his community. "Sure [we were] a ragtag bunch of weekend warriors—which some called us—in the early days," he tells me, "but [we have] evolved to a highly respected organization within the undersea community." Behind Wallace, where the parking lot abuts a patch of dead grass and a stand of droopy maples, one gray-haired

PSUBS member shakes up a can of Miller Genuine Draft and sprays it at another gray-haired PSUBS member.

"Yup," Wallace says, "I have worked in computer engineering for nearly forty years and have no formal training in submarines. But . . . I could start a submarine college today." Though he stresses that he has no regrets about having spent a lot of his free time with his wife and children, he follows with a lament. "It's that life that gets in the way of fabricating or rebuilding a submarine," he says. "If you have to work for a living, have a mortgage, are raising a family . . . then all that can get in the way of submarine dreams and goals." He puts his hands on his hips and surveys the hotel parking lot, the array of different subs, the array of similar-looking men.

Though Wallace calls his obsession with personal submersibles a "passion," he often catches himself when in conversation and tries to present as sober and businesslike and dispassionate. Every time he uses that word, he follows with a sentence attempting to undo it. "It certainly is a passion," he says, "but I am somewhat matter-of-fact about the entire hobby . . . All passions wane on occasion." When I ask him to describe his first dive in his personal sub, he quickly says, "It was exciting!" but then arrests himself and continues in a lower, flatter voice, bordering on the robotic. "No. I did not have a deep emotional response before or after diving my first time."

Wallace presents as a guy's guy. When I remind him of his prior excitability and ask him how it dovetails with no "deep emotional response," he shakes his head before I can finish my sentence. "I had a matter-of-fact response to my first submarine dive." That's it. Case closed.

He rejoins his comrades, and I do my best to eavesdrop inconspicuously on their conversations. "What was down there?" someone asks the owner of the sub *GEN3*, about a previous dive. "Dirt!" he says.

On the lee side of a yellow submarine, someone crows, "It's gonna look totally different. There's gonna be lights on it. There's gonna be a windshield. There's gonna be harpoons!"

"We're gonna get this thing flying," someone else says. "We're gonna chase down some sea life!"

Disconnected exclamations abound, and it's hypnotic, easy to lose oneself in the cacophony.

"Robot claws!"

"Propellers all over the place!"

"Maybe some spearguns!"

The owner of *Persistence*—a sturdy, borborygmic man with an ample chin and robust gray moustache—has cred among the group, as he's fashioned his own merch. He struts around in a yellow T-shirt silkscreened with the image of his vessel, behind which the word PER-SISTENCE is written five times, vertically stacked—a tapered wall of PERSISTENCE. In each subsequent instance, as one reads downward, the letters grow more faded, the last instance of PERSISTENCE barely visible, wispy as a ghost, as if the word itself is sinking, fading into some encompassing sea that, in this case, flows beneath its owner's navel.

"These kinds of things, they, um, they sneak up on you," another p-subber cryptically says of his obsession with the hobby, "like a submarine arriving at somebody's garage." Some of the men laugh knowingly, as if this were some oft-repeated punch line, taking pride in the sort of coded humor meant to mystify any outsider like me.

≈

A group of p-subbers stand in an ellipse at the parking lot's edge. One of them kicks stones at other cars. They crack their beers, and one of them calls, "To Drebbel!" They toast, their cans coming together in an anticlimactic *thunk*. "Drebbel!" they yell.

Culling from Aristotle's writings on the diving bell and Alexander the Great's bloodthirsty usage of the same, and filtered through the subsequent designs that would also later influence the developments of the bathysphere and the bathyscaphe, the Dutch engraver and glassworker turned engineer Cornelius Drebbel, under the patronage of King James I, took off for England in 1604, and by 1620 had completed fabricating for the Crown his "diving boat"—generally credited as the world's first submarine, by modern design standards.

Drebbel improved upon a prototype sketched out by the English mathematician, innkeeper, and celebrated Royal Navy gunner William

Bourne. Bourne conceived of a wooden ship, canopied in a leather shroud waterproofed with animal grease—like plastic-wrapping a Tupperware bowl—that could be rowed underwater with oars. The boat was supposedly able to submerge by using a hand-cranked screw thread to adjust a series of plungers. The plungers would press against a row of goatskin bags lining the sides of the boat, which would increase or decrease the volume of water, in turn affecting the boat's ability to stay afloat or to sink. These goatskins served as the first ballast tanks (the water-filled compartments on a sub that regulate stability and buoyancy, alternately flooding the tanks to sink, and emptying them—often with compressed air—to rise). In his design, though, Bourne included no discernable accommodation for a crew.

Drebbel—pressured to improve on Bourne's design by a regime thirsty for new "devises and strategems for harming of the enemyes by the Grace of God and worke of expert Craftsmen" (or so professed Scottish landowner, inventor, and theologian John "Marvellous Merchiston" Napier in 1596)—adjusted the vessel's shape toward something more bell-like, but kept the fatty leather drape and the propulsion-by-oars. He was concerned with alleviating some of the at-depth pressure that damaged the ears and the sinuses and the brains of the likes of Alexander the Great. Drebbel affixed two long snorkel-like tubes to the sides of the boat, and equipped them with floats to maintain an airflow, however meager, beneath that rank leather blanket. The proto-sub made it from Westminster to Greenwich along the meanders of the River Thames—some nine miles—at an average depth of thirteen feet.

The Crown demanded further improvements, and new models (initially and ominously called "sinking boats") resulted, which begat the subs that further enabled England to ratchet up its reign of colonial terror around the globe, Alexander the Great's barbaric and predatory heir apparent. In 1690, a subsequent model was designed by the same guy who invented the pressure cooker, and it looked exactly like a giant version of the same. Scientists and adventurers later culled from Drebbel's innovations to fabricate the bathysphere (essentially, a steel orb lowered into the sea by a cable from a surface ship), in which early crews claimed to have spotted fish so fantastic that they were later dismissed as illusions born of the fevered mind at depth. Soon thereafter came the bathyscaphe, invented by the Swiss balloonist Auguste Piccard, who used to party with Einstein, and who had already dreamed up a hydrogen-filled gondola. Piccard, who was referred to as a "scientific extremist" obsessed with the uppermost and lowermost atmospheres of our world (he's believed to be the first person to visually glimpse the curvature of the earth), nicknamed the bathyscaphe the "submarine balloon," as it was attached to a "free-floating tank" and had no need for a tether to the surface. It was to float in the water as his balloon did in the air. In a later model of the bathyscaphe, the float chamber was filled with gasoline—which is lighter than water—to boost buoyancy.

The American inventor and darling of George Washington, David Bushnell (who was the first to figure out that gunpowder could ignite underwater, and who designed the first time bomb), used some elements of Drebbel's model to fabricate, in 1776, the *Turtle*—an eggy-looking war sub made of wood, steel, and tar—to attack the very Royal Navy that had enlisted Drebbel all those years ago, as they blockaded New York Harbor during the American Revolutionary War. During the American Civil War, the Confederate Navy constructed three ill-fated subs, one scuttled during an attack, one sank in a storm, and a third—the *H. L. Hunley* (named posthumously for its designer)—after successfully attacking and sinking a Union ship, killing five, strayed too close to that explosion (she was only about twenty feet away from her target) and also sank, killing all eight on board. One of whom, oddly enough, was a Danish sailor who had deserted his freight ship that had landed in Charleston after a voyage from its home base at Dragør Harbor—the same harbor where, some 157 years later, Peter Madsen was brought ashore after deliberately sinking *Nautilus*.

*Hunley* was a hard loss for the Rebels. She was an uncommonly sleek sub for her time, a deadly femme fatale. Her bellows were ample, her propeller slim, her planning system downright stirring. Her hatches were rotund and watertight, her pedestals literal and figurative. Once inside her hull, her angles and measurements made egress difficult. Her crewmembers, according to the journalist Patrik Jonsson, "were forced to crouch, praying and sweating in the lanternlight." But most admired her from afar. The boys nicknamed her *The Porpoise.*

H. L. Hunley himself was an elegant man—a New Orleans lawyer with a head for design, who often stashed a fresh flower at his lapel or behind his ear. His facial hair was neat and sharp. Hunley cofounded the engineering of his namesake sub along with members of the Singer and Whitney families (of the sewing machine and cotton gin empires, respectively). Hunley, though not a member of the Confederate military, was so enamored with his sub that he often insisted on riding within it on missions. He—perhaps beflowered at the chest or temple—was on board when it sank, among the eight who perished. (His body was later recovered and buried beneath a giant magnolia tree in a Charleston cemetery. After a heavy rain, the white petals often cover the headstone, and from a distance it can resemble a stout child draped with a bedsheet, disguised as a ghost, readying its *Boo.*) When the Confederacy demanded that *Hunley* be raised, a morbid *Three Stooges* episode resulted. It was raised, but then dropped, and sank again. It was raised a second time, and a second time dropped. A total of twenty-one sailors died across all three sinkings.

Thereafter, men built subs powered by compressed air and subs powered by steam engines, and internal combustion, and rechargeable batteries, and nuclear power. In the days leading up to World War I, the British were so intimidated by the power of the submarine, they launched a campaign to abolish them worldwide, claiming that they gave whoever wielded them an unfair advantage. They made blockades impossible. "There is no answer to it," one British admiral asserted. "The development of the submarine has ended the possibility of any power in future rating as mistress of the sea."

Other British officers tried to downplay the power of the submarine, which had become somewhat mythological, comparing it to an undersea version of an urban legend. They dismissed them as vessels of "piracy and sabotage," as if the submarine were too ungentlemanly for their preferred versions of war. In the 1920s, they wrote overheated op-eds and published them in such outlets as the Washington *Evening Star* and *The Indianapolis Times* with headlines like "Briton Scoffs at Reputation of Submarine" and "Submarine Value Grossly Overrated."

Such articles—making up in zeal what they lacked in prescience—also took aim at the airplane, which they claimed was "obsolescent... passed the zenith of its usefulness in war."

They were not right. Across World Wars I and II, German U-boat submarines bombed and sank nearly ten thousand ships. Racist op-eds abounded, penned by other British admirals and American colonels, directing their vitriol toward Japan's role in the war. "Japan, under German tutelage, has been quick to plan the construction of gigantic submarines," claimed one. "These submarines will be specially constructed to avert the ravages of bomb or depth charge attacks, and will carry sufficient oil to give them a cruising radius practically around the world."

Such articles became common and often preempted argument. "While the peace propagandists in America are sowing their insidious seeds of discord," said one, "fighting against military training in our public schools and universities, we must take cognizance of what Japan is doing... What does this mean to America? It means that the Yellow Ace of the Imperial Islands will dominate the Pacific."

In the following years, many men built submarines named (*yawn*) *Nautilus*. *Nautilus* is to the sub world what Jennifer and Michael were to the baby namers of the 1980s. These numerous Nautili were used for purposes both benign and atrocious. Mostly—with the benefit of hindsight—atrocious.

Drebbel couldn't have seen all this coming when he was sewing his little goatskin bags. In the style of many well-meaning inventors oblivious to the dark side of wonder, he remained horrified that his device inspired the development of tools of war—subs loaded with mines and gunpowder and torpedoes that often proved more dangerous to their crews than to their enemies. Drebbel retreated and turned his attentions instead to alchemy and painting. He became obsessed with the brilliant red dye carmine, derived from the ground-up exoskeletons of the cochineal beetle, and invented a compound that would allow the dye to absorb more quickly into material. The dye—arguably our world's most beautiful shade of red—was also used to treat our wounds.

As various wars were waged and political allegiances shifted, Drebbel, remembered for his submarine, was occasionally imprisoned. Once freed, he made eyeglasses for those who lived in poverty and built a camera obscura. He built an image projector called the "magic lantern." He fabricated a "perpetual motion clock" driven by fluctuations in temperature and atmospheric pressure. He invented air-conditioning. He cobbled together compound microscopes with which he examined the qualities of droplets of his own blood. He wanted to cool off. He wanted to know who he was, of what he was made. Eventually, he retreated from the public sphere, took up work as the proprietor of an alehouse, and died near penniless in 1633. Three hundred and seventy-nine years later, he was honored on a postage stamp and had a lunar crater named after him.

The p-subbers like to claim Drebbel as one of their own. All early sub builders and captains—from Aristotle and Alexander the Great on down—were p-subbers, the contemporary p-subbers believe. As the sun sets over the parking lot and reflects carmine red from the acrylic lenses of the subs, they claim him.

≈

During their dives in a nearby lake, the men communicate from their submersibles with those onshore via walkie-talkies, and I eavesdrop on their voices—staticky and disembodied and phantasmal—crackling through the speakers, as if transmitted from some transoceanic shortwave signal. If their voices sound as if they're originating from some kind of Great Beyond, it's because they actually are. "Head for the bottom!" one shouts, like a new ghost, uncomfortable in their fresh ephemerality, attempting to navigate limbo for the first time.

Some offer to take me down with them, and each time, my body recoils and I shake my head, though I'm ashamed of myself—ashamed of my lifelong fears of the ocean and deep water; ashamed of my inability to swim and my claustrophobia; ashamed that I'm often obsessed with the objects of my fear. The men laugh at me, and I try to envision what it may be like down there. I envision myself staring out the porthole, some prisoner futilely pining for an escape, and at first it's only dark—a greenish dark. But then another nearby sub turns on its light and I swear I can see a shadow in the murk—some elusive leviathan, some legend made manifest, about to declare itself, rescue itself from the realm of myth and enter, finally, our real world. But, quickly, it seems to lose its power, its ability to surprise and bewilder and exhilarate. "It was what I expected," said Jon Wallace. There are so many shadows down there.

I'm snapped out of the vision by the voices on the walkie-talkies. "Tryna see the bottom seems kinda important!" someone shouts to their fellow passenger as they slam against the lake bed. It sounds as if someone drops his camera. Someone complains that their glasses fell off.

"I don't know what that noise is . . . ," someone else says.

Most, though, in however foreboding fashion, just count off the depth as they descend: eight feet, ten feet, twelve, fourteen . . . as if marking not only their descent but the passage of time. I'm made keenly aware of how much closer we all are to our deaths—which makes the lake look bluer, the clouds and trees and blades of grass reflecting from it more sharp-edged and beautiful in their dancing. The world, during these chimeric countdowns, feels more consequential. The act of diving into this lake seems both a test and a proof of the innate structural inadequacies of such repositories, the incompleteness of any archive that presumes to hold us and our stories—spherical, acrylic, or otherwise—and therefore of any stab at collective memory. The archive—like language, like the submersible, like the drive to make manifest the confines of our obsessions, whether claustrophobic or expansive or both at the same time—is a faulty tool.

I stay on the shore and watch tipsy goateed silver foxes sink below the surface. When they are submersed, I have little to do but wait. I watch the surface of the water smooth, then stare at my feet. I swear it: a fat purple worm emerges from beneath a downed still-green maple leaf and drags itself over my shoe top before disappearing into the earth.

*Special K*'s owner is the first to surface. Through the porthole, I can see him holding the walkie-talkie to his mouth. He has a sweet face and wears thin tortoiseshell glasses and a blue ballcap with GPS coordinates on it that correspond to Saganaga Lake in northern Minnesota, which houses an island called Massacre Island and was once believed to be the place on earth in which the northern lights best reflected. For a while, all *Special K*'s owner does is breathe. Then he finds his voice. "I feel kind of vulnerable," he mutters. His voice echoes from the hull. "Sorry for the reverb," he says, as if to the whole world—lost and found, living and dead—"but we are in a bubble."

# 11

Kim Wall didn't wait long for Peter Madsen to open the door to his Laboratory Room. Moths circled beneath the outdoor lantern. When Madsen heard the knock, he put down some tool he had been fidgeting with. He was often fidgeting with some tool. Maybe he cracked his knuckles, and smoothed his jumpsuit, and bit his lip, and took a deep breath.

His hair was disheveled, as it always was. He was breathing heavily but bore an aura of calm—an aura that he seemed to labor to maintain. He cleared his throat a lot. He laughed at things that weren't funny, and when he laughed, he laughed too loudly. He was performing. It was as if he were watching himself perform, so delighted and so delightful. The smartest guy in the room.

Kim asked questions about rockets, about *Nautilus,* about timelines, about inspiration. According to friends, Kim could be prone to anxiety, which she channeled into her generous reportage; her careful eye; her persistent hustle to land the freelance gigs she cared about and felt were important. She listened to Madsen shifting his phlegm as he tugged at the collar of his jumpsuit, twisted his sleeves, shuffled his feet. When he moved to offer her a cup of tea, he moved jerkily, jumping around like some manic puppet. He likely did as he often did—referred to himself in the third person. Referred to himself as *Captain.* Kim did as she did when reporting from Berlin's Brandenburg Gate, or from a water park in Pyongyang, or from Idi Amin's torture chambers at Mengo Palace. She assessed the place.

The workshop's interior was cavernous—seemingly too large a space to be contained by the iron facade. The walls and ceiling formed one yawning arc and were painted white. Inside, it looked like a missile silo. Littered about were wooden ladders, crates, pallets, fire extinguishers, scaffolding, a refrigerator filled with meals half-eaten and drinks half-drunk. Rocket parts and submarine parts. Against one wall, an obsolete set of three white wooden steps rose to meet only the wall. Kim, of course, took mental notes, began, in her head, to construct the sentences to describe this eccentric's workshop.

Perhaps Madsen took stock of the women in his life—the wife, who lived separately and who would divorce him in the coming months; the exotic dancer he entertained in this workshop years before when fashioning his previous submersible, *Freya*. He had become disturbingly fixated on a moment when the dancer touched his submarine-in-progress, as if it were a part of his own body. "It was so wonderful," Madsen said about that moment, "quite special that she touched it. The place she touched I have always had a special relationship with." Perhaps he thought of his mother, whom he rarely saw during his childhood, and about whom he once said, "[She] was a nice lady, who lived in a nice house, a nice rug and everything. If I made a bike, she was shocked that my fingers were dirty, and in such an environment I could not exist." Perhaps he thought of the women with whom he had been intimate, many of whom, when later asked about his personality, said some variation of "[Peter] definitely can't do *no*."

Days later, when his name and Kim Wall's name were dominating the headlines, he would send a text message to one of his former mistresses that read, "I'm the same Peter I've always been."

Kim sipped tea and started asking Madsen another question. Before she could finish, Madsen interrupted, excitedly inviting Kim to conduct her full interview that very night, within the hour, on board *Nautilus*, as if the idea had just occurred to him. *What a performance. So delighted and so delightful. The smartest guy in the room . . .* Kim likely felt that familiar flutter of anxiety as she smiled, as she felt her story solidifying. She put down her teacup on a crate and texted Ole. Madsen fidgeted with his coveralls, futzed with a screwdriver, stared at his brown shoes.

≈

To what degree is there an overlap between a penchant for claustrophilia (the "abnormal desire for confinement in an enclosed space," i.e., a submersible) and the penchant for committing a murder? And to what degree was Madsen's desire to encase himself in that steel spheroid and sink to depth representative of claustrophilia? A desire to regress, to return to the paradoxical confines and timelessness of the womb state? According to *Psychology Today*, claustrophilia often manifests as a sexual fetish, "an extreme form of bondage whose adherents are aroused by total encasement in tight spaces such as boxes, bags, cages, caskets, and car trunks."

In 2010, the decomposing naked corpse of Gareth Williams, the Welsh mathematician turned James Bond–style superspy with the UK's Secret Intelligence Service (or MI6), who had been investigating the Moscow mafia's involvement in international money-laundering networks, was found tightly zipped up in a red North Face duffel bag, in the tub in his London flat's en suite bathroom. The duffel's zippers were padlocked from the outside. The bathtub's faucet (as are many faucets in the loos of London's marshy Pimlico neighborhood) was leaky and, drop by drop, dripped its water onto the bag, forming the dark red amoeba that had seeped into the portion of the cloth covering Williams's face. The flat was down the street from St. George's Square, where the hot August pigeons flew between the eaves of the white stone town house where Bram Stoker died in 1912, and the public garden's white stone statue of former British parliament member William Huskisson—the first person ever to be run over and killed by a railway car—dressed in a Roman toga.

Williams's flat was set up as an Intelligence Service safe house and was vigilantly surveilled. His death confounded the authorities. The newspapers overused the words *mystery, suspicious,* and *unexplained,* as it was reported that evidence had been tampered with: fingerprints and palm prints, footprints and blood wiped away. In fact, though his body was right there stuffed into that duffel, not a single trace of Williams's DNA was said to have been recovered from the bag's "rim," zippers, or padlock. The key to the padlock was found inside the bag, beneath Williams's body, pressed between two of his ribs, leaving behind a tiny pink key-shaped depression in the skin there. Initially, having concluded that it would have been impossible for Williams to have done this to himself, the police reported that the death was "unnatural and likely to have been criminally mediated." Only later, upon reinvestigation (many details of which were withheld from the public, making the reports cryptic at best), was it found that Williams fancied particular bondage websites and practices, preferred dressing in women's clothing when in private, and had to once solicit the help of his landlady to free himself from the wrist and ankle ligatures with which he had managed to cinch himself to his bedposts. The initial "conclusions" made about his death were subsequently revised, redesignated as "probably an accident."

There were no signs of forced entry or struggle, and it was determined that Williams got into that bag voluntarily and was locked in with the help of a lover, who may have panicked when Williams failed to free himself and quickly succumbed to hypercapnia (too much carbon dioxide in the blood), which is also a common cause of death in deep-sea divers. It was reported that it may have been this mysterious "lover" who scrubbed the scene of evidence and who—though it was mid-August—cranked up the heat in the flat to hasten decomposition, before fleeing.

In his lesser-known novel *The Mystery of the Sea,* Williams's neighbor Bram Stoker wrote, "We were already beginning to feel the chill of the tide . . . our heads were so close to the roof that I felt safe so far as actually drowning or asphyxiation were concerned if the tide did not rise higher . . . And then we began the long, dreary wait for the rising tide. The time seemed endless . . . the water had reached our waists . . . It was a terrible trial to feel the icy, still water creep up, and up, and up. There was not a sound, no drip or ripple of water anywhere; only silence as deadly as death itself."

The name of the train car that mangled and killed William Huskisson, incidentally, was *Rocket.*

≈

Claustrophiliacs explore "the use of space to intensify desire," says Cary Howie, professor of romance studies at Cornell University and author of *Claustrophilia: The Erotics of Enclosure in Medieval Literature.*

Howie stresses that the practice was often pursued by medieval poets and hermit saints seeking overlaps between bodily ecstasy and spiritual ecstasy. "Small spaces from which we cannot escape," Howie says, "make us hyperaware that we have bodies.

"I think your instincts about [the overlaps between] confinement and violence have to be right," Howie tells me, "especially if you're thinking of violence as the explosion of, or maybe the sudden infringement upon, a boundary or surface. Spatial limits have a way of heightening the intensity of the body's own edges. I'm assuming this would be even more the case when you add questions of pressure, which must be a part of the underwater equation. And when you occupy that limited pressurized space with another person, surely all kinds of intensities can be unleashed."

According to Carol Queen, educator and cofounder of San Francisco's Center for Sex and Culture, wedging our bodies into these spaces can produce states of arousal that "stem from a sense of helplessness ... or from altered breathing, which gives a sense of being high ... or from proprioception: the body's experience of itself in space." The thrill depends on the confines of the submarine, and of that submarine's immersion in the deep ocean, a brine we can't normally exist within, or breathe into our bodies, without dying. Only in the deep can some feel this "sense of being high."

Carving out an even smaller space within these other suffocating spaces may ratchet up the high even further. Of course, imposing that confined space within space within space onto another unwilling participant is an act of violence, and sometimes murder; crimes that result in inquisitions, conclusions, funerals and verdicts, tributes and interviews, and the subsequent confinement of a body like Kim Wall's to the water, and a body like Madsen's to a prison cell, where—for one of the only times in Danish history—it will be padlocked for the rest of its life.

# 12

When Kim Wall left Peter Madsen's workshop to return to her farewell party, she closed the door carefully, so as not to tempt the loudness of the corrugated iron. The leaves, in this wind, sounded like insects trying to quiet one another, and Kim walked briskly beneath them, on the cusps of both this interview and this new phase in her life, elated.

Back at the quay, Ole was mingling with about ten of their friends, their food heaped onto white plates. Kim approached her boyfriend from behind. They smiled at each other. They may have laughed a little. They were about to move to China together. Kim felt guilty about the prospect of ditching her own farewell party and wondered aloud to Ole if she should take Madsen up on his interview offer or if she should hang back, celebrate with her friends, and try to organize something for the next day instead. The seagulls swept low over them, eyeing the food, calling out to one another in warning. Ole convinced Kim to go ahead with the interview. She had been waiting for this for so long, Ole said, and it was to be on board his homemade submersible no less. Plus, Ole said, she could easily peddle the resulting article to some big-time American magazine. She had already made contacts at numerous publications and recently received some eager email from *Wired*. Kim asked Ole if he wanted to join her, and, sure, he wanted to check out the submarine, but he shrugged, and made a rueful face, and gestured to their guests. It shouldn't be more than a couple of hours, Kim said, and Ole looked at their friends, the wine in their cups, their jacket sleeves rustling in the wind, their faces so happy, and said that it wasn't too late, said he was sure everybody would still be here.

Kim told everyone where she was going, and they made Os with their mouths, and congratulated her, and told her of course, of course they would wait until she returned. She let her hair down, gathered up the loose strands, and retied it. *Potato,* the papers would say. Madsen's submarine was docked only a few hundred meters away, and Kim waved to her friends, waved to Ole.

The 34 muscles and 101 ligaments, tendons, blood vessels, and nerves wriggled like fish in the aquarium of her hand, depended on one another, communicated with one another, exerted so much energy so quietly that Kim herself did not recognize their toil. The thenar eminence danced with the hypothenar eminence as the lumbricals wormed their way against the carpal tunnel. The ulnar nerve leapt through Guyon's canal and branched out toward the median nerve, the radial nerve. The bursae sacs secreted their lubricating fluid. The soft tissue softened. Her hand felt no pain. Blood rushed in from the thumb and the pinky, these two streams meeting in the middle. Kim had bent and stretched the fingers of this one hand over ten million times in her life so far. They had felt and perceived so many things, moved through so many different qualities of air. The seventeen thousand touch receptors blossomed in Kim's palm, tingling with the pressure, movement, and vibration of the external world. Our palms and our fingertips are so sensitive that biologists have dubbed them our "antennae."

The wave has been called by the American Psychological Association one of the more essential components of human language, nonverbally acknowledging the presence of another, greeting, bidding farewell, denying, calling for a merciful silence. We remain mystified as to the origin of the wave and its original uses and meanings. Maybe it began in the eighteenth century as a salute among soldiers and the aristocracy and was then co-opted by the proletariat, at first as a way to satirize and ridicule the nobility. Among knights, the salute replaced the more unwieldy gesture of removing one's helmet to show one's identity, demonstrate vulnerability, and that they came in peace, meant no harm, carried no weapon. But according to American Sign Language, the deaf population had already been using the wave (sometimes accompanied with a handkerchief) as a way of calling someone's attention, demonstrating approval, applauding, saying hello, saying goodbye. Different variations of the wave could, of course, mean different things. Stilling the wrist and wiggling the fingers could demonstrate a flirtatiousness. A grand gesture of the arm, accompanied by a single wrist-flick, could demonstrate a dismissiveness. A slow, metronomic gesture could be plaintive, longing, elegiac, heartsick.

In the New Testament, while crossing the turbulent Sea of Galilee, Christ waves at the sky in order to calm the tempest. In Hindu, Buddhist, Jain, and Sikh stories, the wave takes the form of the *abhayamudrā*, an indication of calm in the face of potential danger, meant to assuage and protect both the waver and those who behold the wave; it has the power to make rabid elephants fall asleep and make warlords lay down their weapons. In the ancient Greek Eleusinian Mysteries, the wave was deployed to halt the rebirth of Persephone and curse the forthcoming harvest. Some scholars assert that, in the Greek legend of Orpheus and Eurydice, it was not Orpheus's loss of faith in the gods as demonstrated by his turning around to see his beloved near the exit of Hades that doomed Eurydice to everlasting internment in the underworld and cemented the lovers' separation, but the fact that, when Orpheus turned around, Eurydice lifted her arm and waved to him. It was the wave that doomed them, "the upraised arm of a woman," according to the art scholar Erika Doss, being perceived (and subsequently punished) as a "gesture of emotional frivolity."

But even in the making, and subsequent dissecting, interpreting, and remaking of these myths, "the mythographer," according to the writer Roberto Calasso, "lives in a permanent state of chronological vertigo, which he pretends to want to resolve . . . The mythical gesture is a wave which, as it breaks, assumes a shape, the way dice form a number when we toss them. But, as the wave withdraws, the unvanquished complications swell in the undertow, and likewise the muddle from which the next mythical gesture will be formed." The wave, in its perfect inadequacy to encapsulate and ground the nuances of communication, experience, and feeling, symbolically drives, it seems, the evolution of the stories we tell ourselves—dizzily—about ourselves.

In Nigeria today, so as not to be offensive, the wave is performed with the palm hidden by the fingers. Grecians wave with the back of the hand, and South Koreans wave vertically, the palm downturned. In

China, the wave is mostly reserved for women greeting other women. Cognitive and experimental psychologists postulate that the wave may be innate, even if its various meanings, uses, and interpretations are learned. Preverbal infants typically wave as a demonstration of their first visible sign of conscious communication. This occurs even when an infant has never observed another person waving. Human fetuses have been observed waving while in the womb, stirring the amniotic fluid with their hands, essentially making waves in the water. We've named, after all, our most primary of gestures and stabs at communication after the tidal pull on our oceans, the frolic of the sea, the undulous agitation of the amniotic.

Our waves goodbye borrow so much from the water or have been gifted so much by it. Kim Wall waved. The wave crests and falls, transmits energy, and not matter; transmits something like affection from one source to another. The wave says, I cherish you, I will miss you. The wave knows: it may not return; this transmission of energy may be, this time, one-way, a detour, a dead end. The wave can rise only after first sinking.

# 13

ON FEBRUARY 9, 2001, off the coast of Oahu, *Ehime Maru*, a Japanese high school fisheries training ship, on which students were busy taking hands-on classes in long-line tuna fishing, oceanography, marine engineering, and maritime navigation, was struck by the American nuclear sub USS *Greeneville* on its hasty rise from the depths—the sub accelerated from a depth of four hundred feet to the surface in only a few seconds. In less than ten minutes, *Ehime Maru* sank, killing nine of the thirty-five people on board: three crewmembers, two teachers, and four high school students. The U.S. naval officers on board *Greeneville* blamed the "waves" for their aloofness in the aftermath of the crash, during which they stood by watching *Ehime Maru* go down and made no effort to help. Even after the surviving teachers and students, of their own volition, deployed and scrambled onto their lifeboats, the U.S. Navy made no offer of aid.

In the immediate aftermath, *Greeneville*'s captain, Commander Scott Waddle, refused to apologize, concerned more with downplaying the gravity of the crash to the sixteen civilian Distinguished Visitors (DVs)—a group of wealthy CEOs and their spouses—who had been invited on board his vessel. Regarding *Greeneville*'s rapid rise to the surface, it was later reported that Waddle was "frustrated that he couldn't start the maneuvers right away" and dismissed some of the "equipment preparations" that would cause any delay. His justification, later recorded in the *U.S. Navy's Report of Proceedings: Marine Accident Brief*, was that he was performing "drastic, high-speed full-rudder" turns and "rapid up-and-down-movements," and that the wealthy civilian DVs on board "were loving it." He even let a couple of the civilian guests perform the "main ballast blow," which rocketed the sub toward the surface. "Jesus!" Waddle exclaimed when *Greeneville* struck *Ehime Maru*. "What the hell was that?"

Eventually, the Naval Board conducted a Court of Inquiry, during which they found Waddle (who neglected to perform adequate sonar sweeps and a periscope search before rapidly surfacing, and who, prior to setting sail, failed to fix a broken video monitor that would have communicated crucial sonar information to the officer of the deck) culpable for the accident and the subsequent deaths and forced him to retire (with, inexplicably, an honorable discharge). It was up to a team of Japanese divers to recover the bodies of those dead students, teachers, and crewmembers. They found the remains of eight of the nine.

Less than two years later, Waddle—as Cmdr. Scott Waddle (Ret.)—authored, for a sizable advance, a book (which is typically categorized as "Christian" by booksellers) about the incident. He cast himself as a tortured motivational figure who—via his religious conviction—found his innate integrity in accountability, therefore providing "an inspiring challenge to anyone who is facing difficult choices in any area of life," a tagline that is woefully vague in conjunction with the specific details of the crash. The "area of life" in question, of course, was the control room of that submerged submarine, and the "difficult choice"—however suffused with the spirit of Christ Waddle wants to make it after the fact—was to persist in enacting the reckless maneuvering that seemed to be getting off the speed-freak CEOs.

Waddle's book is called *The Right Thing,* a title that, on the book's cover, is rendered in glossy gold letters looming over the head and shoulders of a fully uniformed Scott Waddle wearing on his face his best expression of contemplation that his targeted readership will no doubt mistake for "character." The title is, in fact, part of a longer sentence, the beginning of which precedes it on the cover in smaller white letters, reading, "The untold story of the deadly collision of the nuclear submarine USS *Greeneville* with a Japanese fishing vessel, and one man's courageous decision to do . . . *The Right Thing*." The book is endorsed by then-president George W. Bush, who calls Waddle a "fine American patriot" and includes such chapters as "Showtime," "Spiritual Strength," and (yes) "It's Hard to Say 'I'm Sorry.'"

In the final chapter, Waddle writes, "People often ask me, 'Scott, what got you through this ordeal? Other people have crumbled in the face of far less pressure. How have you survived? How have you been so strong?' My answer is simple: My friends, my family, my faith in God . . . We will never be able to repay the many friends who helped us in tangible ways and the myriad people who prayed for us . . . God showed me his love through strangers, through their expressions of love, their unmerited gifts, and their acts of kindness. The United States is an incredible nation, and the American people are truly amazing."

Ryosuke Terata, whose seventeen-year-old son perished in the crash, addressed Waddle in an interview with the Kyodo news service, stating, "In Japan [you] would be fired and indicted on charges such as professional negligence resulting in death . . . I don't know if I will ever be able to forgive you," a sentiment to which Waddle responds, in his book, "I still have many unanswered questions. But they don't matter anymore. This accident has strengthened my faith . . . Oh, I still hold my head up high because I know that I am a child of God."

Today, Waddle makes his money as an "inspirational speaker and leader with uncompromising ethical standards" who commands an uncompromising speaking fee of $14,999 plus expenses per event, and is "perfect for: Associations, Banquets, Closing Sessions, Corporations, Entrepreneurs, Entry Level, Management, Men's Groups, Opening Sessions, Sales, Senior Executives, Spouses, and Women's Groups," who will "1) Learn to pursuit [sic] of integrity against all odds 2) Learn an inspiring challenge to anyone facing difficult choices in life 3) Learn that Failure Is Not Final."

In a 2021 interview with *Hawai'i News Now,* commemorating the twentieth anniversary of the *Ehime Maru* tragedy, Waddle, his hair gray and the skin at his throat loose and shuddery, tugs at the collar of his blue-and-white plaid shirt. He admits that his guilt over the incident grew over time to the point at which he nearly enacted a strange penance. "I thought about going in and killing my daughter Ashley—then thirteen—and my wife. They were in bed sleeping," Waddle says, adjusting his rear on the cushions of a khaki couch. It's unclear what Waddle wants out of this admission—whether it's meant to further absolve him of his guilt, testify to his "tortured" state of mind at the time, or reveal that he's the sort of man who would make up for his mistakes by enacting violence on the bodies of his female family members. "It would be oddly horrible," he says, and shrugs, "but then I'd kill myself."

When he admits this, his hair is neatly combed, and his shirt doesn't have a single crease, and his voice seems bored, as if he's repeated this line—perhaps to misguided applause and fat paychecks—at a multitude of motivational speaking events. In the overbright videos of these events, Waddle still self-identifies as a "warrior." He calls the accident that killed so many a "setback" and a "disappointment" and claims that "those disappointments don't necessarily have to define who you are as an individual." The jokes Waddle tells are not funny, but his audience always laughs.

Waddle also newly blames a public affairs officer for his lack of apology in 2001 and claims that the public perception of his lack of remorse added to his own pain. When addressing the incident, he names USS *Greeneville* and speaks of it reverently, but he rarely intones the name *Ehime Maru*. Instead, he refers to it as "a Japanese fishing training ship" or "it." Perhaps he never learned how to pronounce the name correctly, or perhaps it's still too painful for him to say. In one speaking session, he actually places some of the blame for the accident on *Ehime Maru*, claiming that it was "bearing down on our position."

When he speaks of the collision at these events, he usually underscores it with melodramatic music typical of some 1980s action movie. The music swells when Waddle says, "I said a prayer for the courageous. I said, 'Thank you, God. Thank you, Jesus.' I could live with the fact that their vessel was sunk... [but] it was as if somebody had taken their fist, jammed it into my chest and ripped out my heart. I wanted to scream, 'God, no! Anything but this, God, please no, God!'... How embarrassed I was!" Still, he had the presence of mind to instruct all on board *Greeneville* that day "not to embellish their story."

At a 2021 memorial for the deceased at the Uwajima Fisheries High School in southwestern Japan, surviving family members lay white flowers at the base of a black marble monument, and ring a bell—retrieved from the sunken ship—nine times, one for each of the victims. At 8:43 a.m., the time in Japan when the collision occurred, participants have a moment of silence. When they speak again, one man, fighting back tears, says, "This accident has caused great anger and deep sorrow that will never be healed." The father of one of the students who died aboard *Ehime Maru* has no time for what he believes are Waddle's empty words and performances. "Why is he [saying all of this] after such a long time?" The families who lost their children are still mystified that Waddle never made a face-to-face apology. "I should have tried harder to [do that]," Waddle admits, "but did not."

"[But] when you're a leader," Waddle says, starched, combed, perfectly rehearsed, "responsibility and accountability are absolute." He enumerates platitudes with his fingers, starting with his thumb. "Keep your character and integrity intact," he begins, before detouring into a deifying account of an early-twentieth-century admiral named Hyman G. Rickover ("Father of the Nuclear Navy"), who was one of the officers on USS *Nautilus* (SSN-571), the world's first operational nuclear-powered submarine. Rickover became famous among the officers for being a proto–motivational speaker of sorts, preaching his "Womb-to-Tomb" approach to management, and Waddle, continuing to count off the maxims on his fingers, cribs a few of Rickover's "Seven Rules of Success."

"Practice continuous improvement," Waddle says, "establish quality supervision; respect the dangers you face; train, train, and train; learn from past mistakes."

Rickover, it stands to mention, did not heed the final maxim when he was asked, in 1982, to comment on his own culpability for creating a nuclear navy, and if he had any regrets: "I do not have regrets. What I accomplished was approved by Congress—which represents our people. Why should I regret that? My assigned responsibility was to develop our nuclear navy. I managed to accomplish this."

Waddle's idolization of Rickover is further mystifying given that Rickover was infamously implicated in a scandal involving the cover-up of structural welding flaws in submarines via falsified inspection records. Over a sixteen-year period, Rickover had apparently accepted gifts from the submarine-building corporation General Dynamics, ranging from jewelry and furniture to exotic knives. Rickover regifted most of these to politicians to gain their favor. In response to these allegations, Rickover simply said, "My conscience is clear," the sort of glib dismissal that perhaps inspired Waddle's own response to the *Ehime Maru* tragedy. Rickover was forced to retire by then–secretary of the navy John Lehman, who accused the admiral of creating a cult of sycophants that ultimately damaged the Navy and its reputation, what became referred to as the Rickover Problem. A number of submariners who worked with him later referred to Rickover as a "tyrant."

But Waddle, stars in his eyes, has not left the cult: "You see, Rickover was a busy man, and he also knew that if he could put you in uncomfortable positions . . . he could truly determine the measure of that individual."

Waddle goes on to tout his own intellect, common sense, and former status as an Eagle Scout by citing, rapid-fire and with an uncontainable grin on his face, old praise the mighty Rickover long ago bestowed upon him. He appears as if about to hug himself. He again counts off on his fingers some of what he believes to be his characteristics, a fat gold ring on the final finger. "Integrity, accountability, responsibility," he says, and his jowl shudders. "Uncompromising integrity breeds trust. You can put your life in the hand of another individual and know you'll be okay because he will do the right thing. I didn't say *he or she* because in the submarine force, it's still one of the last bastions where only men serve."

≈

Shanee Stopnitzky, one of the few female members of the personal submersible community, feels that such cloak-and-dagger responses to submarine disasters are, in part, the product of a culture of toxic masculinity that's been allowed to thrive on subs—a culture many men (and governments) have an interest in protecting.

"So, it seems like the submersible community is pretty much dominated—" I begin our conversation.

"By a bunch of dudes," Stopnitzky interjects. "Yeah," she says, "it's a lot of dudes . . . I stopped working on the [*Pisces VI* submarine] project because it was not a good working environment. It wasn't feeling fun anymore. It was feeling like a drag. I think the misogyny happens . . . it discourages women from going down this path. I've always had a strong personality and, you know, I'm quite smart. But I know if I had a less forceful personality, it would be very easy to get discouraged, and people would take me less seriously."

Stopnitzky names a few specific male members of the community who "hate women . . . They've treated a handful of people very badly . . . And they've done really harmful things to other people who are really important to me. The stuff they did to me was pretty minor, but, you know, I kinda saw the writing on the wall . . . They're these right-wing, very narrow-focused, kind of, nerds . . . [They] have issues with women, and feel they have been victimized by women, and, just, hate women . . . so, sometimes it can be a violent environment, yeah."

Speaking of the misogynistic faction of the community, Stopnitzky mentions that some see the submersible as a "male" space.

Only in 2010 did the U.S. Navy lift the "ban on women on submarines" that Waddle referred to so reverently, prompting the question—why reserve this male-only space specifically on submarines?

Naval and submariner identity had, since its inception, depended on the foundation of a "premodern hierarchy of master-subordinate relationships in all-male groups," according to Simon J. Bronner, professor of gender studies and folklore, of the National Sexuality Resource Center. And the more the "crude activities [were] hidden from public view," as in a submarine, the more the aggressiveness was allowed to thrive and, under such protection, grow more extreme. Some naval submariners have been known to engage in a "shellback" initiation ceremony. Younger initiates (or "pollywogs") are taken on a voyage by their superiors across the equator and endure a secret and sacrificial hazing in order to leave their pollywog status behind and to become shellbacks and therefore better equipped to handle the tantrums of the sea.

Throughout the trajectory of these ceremonies, the pollywogs had to endure brutal beatings with wooden planks and being stripped naked and prodded with a Devil's Tongue (an electrified strip of metal). Sometimes, the men are made to dress in drag and don bikinis and participate in beauty pageants judged by King Neptune, Davy Jones, and her highness Amphitrite (played by the vessel's highest-ranking officers), after which they are made to endure suffocation rituals (called "equatorial baptisms"). These rituals sometimes involve tying the sailors down on their backs, plugging their noses, and filling their mouths with raw eggs or hot sauce or aftershave lotion; sometimes involve being locked in a "water coffin" filled with seawater and fluorescent green sodium salt. At the conclusion of these so-called endurance tests, the pollywogs often seal their initiations by being forced to kiss and/or lick clean the axle-grease-painted belly of the royal sea-baby and the feet of the sea-hag (each embodied by a superior officer), before being allowed to kiss the ring of King Neptune (another superior officer). According to tradition, the officer "playing" the royal baby wears a diaper, and the axle grease is sometimes substituted with a variety of substances ranging from cooking oil to shaving cream to mustard to eggs to cherries to oysters.

Even the various sub-honors sound disturbingly cultish, from the Top-Secret Shellback (for submariners who have crossed the equator at a classified degree of longitude) to the Golden Shellback (for those who have crossed at the 180th meridian), to the Century Club, the Realm of the Czars, the Order of the Ditch, the Order of the Deep, and the Domain of the Golden Dragon. Men, right?

Tradition dictated that the men out at and under the sea were sufficiently "liberated" to play a multitude of roles—they could ridicule that which they were also permitted to briefly embody, and embody with verve and kink and violence. They could be the Macho Man, but also the Disciplinarian Mother, the King *and* the Queen, the Captain *and* the Madonna, the Plebe *and* the Whore. They could, according to Bronner, simultaneously release themselves from and newly tether themselves to "the umbilical cord of maternal constraint." They had their cake (or belly-oyster) and ate it too.

In the 1996 Supreme Court case *United States v. Virginia*, the Virginia Military Institute battled to keep women out of submarines on the grounds that their presence would decimate these "time-honored" "rituals and traditions" architected specifically for an "all-male institution."

It's not hard to impose a Freudian analytical framework over and onto the sub and those who are drawn to it even in non-naval contexts. Many, Peter Madsen included, have seen the sub as a safe haven from surface mores and government strictures. The sub, for these men, is a place of ultimate control, where they can exact a brief lordship over their own womb-like world—a womb created by them, for them, and that gives back by incubating only them. In here, they don't have to share their resources. In here, what they say and do, goes. In psychoanalytic theory, the so-called womb fantasy is a regressive desire to both return to and subsequently *exist* in the mother's womb, "usually expressed in symbolic form, such as living underwater or being alone in a cavern." In many of these instances, the world—or the *surface* world—seems too forbidding, too thwarting of their desires. A retreat to a womb-like space not only provides comfort and a restoration of agency, but also prepares them for a symbolic "rebirth" into a fresh world wherein their desires would be sated and rewarded, and they would become a person who attained the power and prestige they feel they deserve.

Of course, when we self-institutionalize in this way, we often—as has been the case with some who've served long prison sentences—become dependent on the confines of the institution and can feel threatened if an outside element, a reminder of that larger, forbidding world, slips in through the cracks. To some, this can feel like an invasion—the monster crawling under their tightly pulled blanket.

Harrowingly, some psychologists believe that this aspect of the womb fantasy—"the desire to completely escape your own self and exist in a state other than your usual one"—is similar to the sort of "regenerative dissociation" that also manifests in those who commit murder. The psychologist Andrew K. Moskowitz stresses that this "dissociation was found to be linked to violence in a wide range of populations . . . and was often expressed in the violent act itself, in the form of depersonalization." Moskowitz found links "between 'trait' dissociation and violence—evidence that long-term dissociative processes [such as, one may argue, prolonged periods spent in a submersible at depth] may predispose vulnerable individuals to violent behavior, and even homicide . . . [including] fantasy-driven violence, often accompanied by some form of identity alteration."

It's easier to consider hurting another from a distance—firebombing a village from an airplane, torpedoing an island from the depths. We can more easily dissociate from the lives we are destroying. Even benignly gazing down at a city street from the vantage of a skyscraper's observation deck, we often compare all those people down there—going about their lives, harboring their small worries and disappointments and joys and heartbreaks—to ants scurrying about; ants, who are so easily exterminated with bait and traps and the soles of our shoes. Peering up to or considering the surface from the deep sea is the gazing down from the skyscraper in the inverse. The people there are fathoms away—a thousand feet away, two thousand feet away, so small and scurrying. "Look on up from the bottom," as the old song goes. It's easier from this vantage to reduce the intricacies of the human experience to that of rats in their maze. It's easier to impose upon them the dispassionate and godlike gaze of the scientific observer.

"Distance is integral to dehumanization and the violence that comes with it," says the scholar Hannah Arendt. "Dehumanization is a result of the rejection from a community and the benefits it offers, primarily its social bonds and the entailing legal protections."

"The outsiders are distanced and thus reduced to 'bare life,'" posits the political philosopher Giorgio Agamben, "susceptible to a meaningless death."

Even when a so-called insider and outsider share close quarters—like the sub itself—this conceptual distance can remain for those who spend inordinate amounts of time cloistered at depth, othering those who mill about on the surface. According to a 2016 psychological study conducted at the University of Maryland, a "psychological distance" can increase one's "willingness to engage in ideologically based violence ... A functional MRI brain imaging study found that participants who followed prompts to disengage or engage from a victim—intentionally increasing and decreasing their social distance to the victim by thinking of the victim as a doll or conversely imagining the victim as themselves—found that these engagement tasks altered their emotional response ... Those who perceived a greater psychological distance from the victim had less intense emotional responses."

Even within the tight confines of the sub, one can see another person at this conceptual distance, as a mere surface dweller, an invader of one's manageable, wombic sub-world. An ant costumed in humanity. And when ants invade the things we call our dwelling spaces—our homes, our domains—our default is so often to open the cabinet, lay out the traps.

And if the violence is perpetuated at depth, then it can seem to occur at an insurmountable distance from the surface and from the sorts of people—beholden to their surface systems—who would condemn such violence and punish it. The sub can become a safe space wherein an atrocious man can enact his violent tendencies without those inconvenient surface associations or consequences. A last bastion, cushioned by the belly of the sea.

The critic John T. Irwin states that "the literal return of an adult body to the womb, its total reincorporation, would necessarily involve violence to the mother's body or the son's, and ... thus there is a sadomasochistic component to the ... fantasy of a total return." In this way, the womb, and the submarine, can become both mother and lover, a dually satisfying, dual companion to the sadomasochist voyager on his journey toward rebirth.

How might such "voyagers" react if other women—real women who live on the surface of the earth—enter their "safe space"? Would these women—who are neither these dudes' mothers nor lovers—be perceived as threats to the dynamic, the stasis, the "journey"? Does the fantasy become perforated, reminders of the outside, upper world slipping through the cracks? If these men are asked to embrace a woman as a colleague, what happens to the mother and the lover, and womb-cum-submarine-as-incubatory-pedestal? In what ways—whether benign or extreme—may some feel that this "outsider" needs to be targeted, assaulted, expunged?

Again, only in 2010 did the U.S. Navy lift "its ban on women on submarines," prompting the American Legion to launch a survey, asking the question "Is it a good idea to allow women to serve on US Navy submarines?" The Legion itself preempted a "No" vote with the reasoning: "No, because such a decision opens the door to too many opportunities to unwelcome fraternization"—essentially a version of the "men won't be able to control themselves" argument in such tight quarters.

"Let them have their own subs!" cried one male officer. "Out to sea for months on end and we all know that we Navy men are freaking hound dogs. I simply worry." "I know that it is not a good idea," said another. "I agree that both sexes should be allowed to serve their country, but not together in such close quarters on a submarine." "Having talked with many retired Navy officers and enlisted personnel," said another, "they say NO to the ladies. Seems the girls cause so much drama and stress that they literally bring down the combat readiness. It is time to stop all this political correctness in the military." "My answer is HELL NO!" said another. "Long story short, when equal numbers of men AND women are defending, and dying, for their country on the front lines, i.e. Afgkanistan [sic], then and only then should the military entertain such silly ideas as 'women serving on submarines.' Oh, and here's an idea. Let women to sign up to serve on a sumbarine AFTER they register to sign up for the draft!!!"

Others cite "hygiene issues," which uncomfortably harkens back to the biblical conjecture (one of the early forces cementing the institutionalization of misogyny) that women were unclean due to their periods and should occupy separate sleeping quarters from men. (Early human habitations, following biblical schematics, were often built with two adjoining mounds, two separate dwelling spaces.)

Still others argue it would be too expensive to convert the submarines to accommodate female berths. Others fret that if a woman should "somehow" get pregnant, the pressure in the submarine at depth can be a hazard to the fetus and possibly cause birth defects. And if a pregnant officer needed emergency medical attention, a mid-ocean evacuation would be too expensive. A disproportionate number of folks in this camp stitch their worries to this hypothetical unborn fetus. Despite the fact that the Navy frequently sees emergency evacuations of male officers suffering from, say, appendicitis or a heart condition, and despite the fact that less than half of one percent of female naval officers have "somehow" become pregnant on surface ships, these arguments persist.

Among the most vocal of the detractors are the wives of male naval officers. "They do not think it's a good idea," said Linda Cagle, who owns a restaurant at the Kings Bay Naval Submarine Base on the Georgia–Florida border. "They do not want women on submarines with their husbands or their boyfriends." Jason Mason, a retired officer who served on ballistic missile submarines, feels that sexual harassment on "an integrated" sub is inevitable. "You cannot close the hatch on a submarine, submerge, and tell the crewmembers, 'Don't act human.' We can have idealistic expectations, but we must live in a realistic world."

And in this realistic world, sadly, the submarine remains a generator of violence against women, and said violence is repeatedly and institutionally shrugged off as an innate aspect of male "humanity." You apparently cannot close the hatch on an integrated submarine, submerge, and tell the crewmembers, "Don't viciously harass female officers and repeatedly videotape them while they are undressing or showering on board," which is what happened in 2014 on USS *Wyoming*, one of the first U.S. Navy submarines to welcome women.

When the men didn't want to get caught filming the women directly, they installed hidden cameras in the bathroom facilities. The videos were distributed among the male crewmembers. Such harassment occurred damn near constantly over the course of an entire year. From the get-go, one female officer said, the male officers took an "intense interest" in the women serving on board *Wyoming* (which is based in Kings Bay, where all those Navy wives were wringing their hands, aware of the sorts of men they had married). "[I tried] to stop the [male] submariners from being assholes," she said, "and to get the women to stop crying."

Senior officers repeatedly pledged, and repeatedly failed, to prevent further such incidents, blaming not these men but instead the "conditions" of serving on a sub. One male officer offered this as a solution: "Have all females who want to serve on subs sign a disclaimer that they understand and accept all ramifications of living with men. It'll take a special kind of woman to do that. And then let the good times roll."

Even the Naval Criminal Investigative Service, which investigated the case, refused to classify the incident as "sexual harassment" or "criminal activity," but rather as a "privacy violation." And once again, the language used by senior officers placed the onus on the women to decide whether or not serving on a submarine is "for them." "I just want women, who are all starry-eyed about this," said one officer, now retired, "to know what they're getting themselves into."

Can one really close the hatch and reasonably expect the enlisted men *not* to create what they themselves called a "Rape List" of the female enlistees, as had happened in 2019 on USS *Florida*, also based in Kings Bay, and only the second submarine to welcome female officers? Once discovered, the "Navy leaders" on board failed to address the concerns of the women who were victimized and continued to allow the "lewd and sexist" behavior, which was "tolerated up and down the chain of command," according to the seventy-four-page investigation into the misconduct. The list ranked the female officers according to their body parts, desirability, and "rape-ability" using a star system, followed by comments that included graphic descriptions of "aggressive sexual activity."

"Significant numbers of females became concerned for their safety," said Rear Admiral Jeff Jablon, who admitted that the commanding officer of *Florida* failed to open a formal investigation and concealed the Rape List's existence from his superiors (even though the list was regularly updated, printed, distributed, reprinted, redistributed, and stored on the submarine's official computer network). *Florida*'s commander went so far as to try to prevent, and then deliberately "slow down," a probe following the female officers' speaking out, dismissing the Rape List as "only a piece of paper." Another commanding officer demanded that the female officers "suck it up and not add to the drama."

According to the psychoanalyst Tony Hacker, some men may pursue the "comfort" of the submarine if their sexual desires and advances are thwarted on the surface; if they can't cross that boundary on the cold, hard earth, some may seek an alternative "permeability," a controllable device that will allow them to pass through a different kind of boundary, one that, with the aid of the sub, is more easily "solvable." If you live through the dive, the sea hasn't rejected you. It has embraced you. The sub, too, has embraced you. You have proved your mettle and your manhood. If the sub or the sea rejects you, they kill you, which means you don't have to live with the rejection.

For some men, the presence of a woman on the submarine not

only conjures their inadequacies, but also reminds them that the sub itself—the device that they only thought they controlled and in which they could pilot themselves toward some easy understanding and satisfaction—isn't so different from land: rife with whimsy, chance, perplexity. In the face of this reminder, the stories some of these men told themselves about themselves fell away, revealed themselves as flimsy. When someone of this sort is stripped of their armor, they may try to put it back on by any means necessary, including violent ones. For a certain type of fixated, infantile, regressive man—as well as a system—misogyny and assault become an awful stab at self-preservation.

After the *Florida* case went public, other women reported being hazed in abusive and derogatory ways and with "undeniable malice." Earlier stories surfaced regarding "sexually charged rituals" involving a gauntlet wherein the male enlistees would line up on either side of the submarine's corridor, and the female enlistees would have to travel down the center, with the men "grabbing their breasts, buttocks, and crotch area as they tried to make their way down the hallway." Other stories surfaced of "butt biting" ceremonies, wherein women had to crawl through such a gauntlet. One female officer reported that women were pressured by their superiors not to complain about such incidents: "The label of 'whiny bitch' was especially to be avoided or else, 'your life is going to be hell.'"

But even after these allegations were made, one top officer and defender of such naval "traditions" went so far as to say that these misogynistic rituals were more "for building men rather than attacking women." Another "top admiral" argued that this incident was "an isolated event" and not indicative of the systemic misogyny that banned women from submarines until 2010 in the first place. "[This incident is] not at all reflective of the overall outstanding performance and behavior of our submariners force-wide," he said—seemingly unaware of the misconduct on *Wyoming*, the crotch-grabbing and the butt-biting, and the comment made by that member of the Naval Criminal Investigative Service about all the "starry-eyed" women who should be more aware of what they're getting into—before admitting, "I cannot guarantee that an incident such as this will never happen again."

In lieu of such a guarantee, some female officers have posited sage and sober arguments for a culture of basic respect. "We are supposed to be PROUD of our military, right?" one female officer asked. "Don't we call them military professionals? Shouldn't PROFESSIONAL people be able to keep it in their pants? It sickens me that we take it for granted that our military could not cope with men and women living and working together. This isn't high school. It's the military. How can we expect our boys and girls to make life-and-death decisions and bear the realities of war and yet claim they couldn't handle living together in close quarters?"

Though banned on U.S. Navy submarines until 2010, women long have been building and designing submarines and other vessels for the Navy. Following the attack on Pearl Harbor in 1941, after thousands of men were compelled to enlist in the armed forces, the Boston Navy Yard—the primary site for shipbuilding and repairing in the United States—solicited the help of the demographics of the local populace they had refused to hire during peacetime, primarily women, Black people, and the disabled. In 1943, over eight thousand women worked as shipbuilders in the yard—some promoted from clerical positions—as welders, electricians, pipefitters, painters, and machinists. Grandmothers worked alongside high schoolers, college graduates, and married women whose husbands were recently deployed. As they built, they lived in the berths on their vessels-in-progress, berths that, once they finished the job, they would be forbidden from inhabiting for the next sixty-seven years.

In the sepia newsprint photos, they wear resolute but sly expressions that barely conceal what appears to be joy. They worked long hours. At night, from a great height, its silver and orange lights winking, the Boston Navy Yard looked like some arachnid, stretching its legs out onto the harbor. The lights danced on the water, and the sounds of the machines the women operated—from on high—were barely audible. After they finished a job, and a submarine was completed, they celebrated. For these parties, they changed out of their work clothes and tied their hair into kerchiefs, put on necklaces and drank wine, sang and danced. When one such celebration coincided with Easter, they set up a long table in the yard's structural shop, draped it in paper streamers, and lined it with platters of sweets. Four tall candles ran down the table's center, skinny tendrils of smoke wafting over their faces. They decorated the room with rabbits—ceramic rabbits, clay rabbits, rabbits made of glass and papier-mâché. Celebrating, they laughed with one another. They whispered into one another's ears. They wrapped their arms around each other's shoulders and waists. Some of them held hands.

Two years later, in 1945, once they had constructed the vessels the war effort required—the effort that ensured the U.S. Navy's victories in both the Atlantic and the Pacific—the female crew was slowly laid off from the yard, puncturing this tight-knit community and erecting fresh barriers to similar sorts of work. The crew was left to carry their experiences back into their peacetime lives—an oppressive peace. And these experiences braided with a new frustration, which they also carried into their homes, their worlds, and into the future. Members of the crew staged protests and launched progressive campaigns that paved the way for a new women's movement, fighting to change the landscape of the American workforce and laying the foundation—however infuriatingly slowly—for women to pursue careers in any field they wished. Some of these women had children, and some were daughters, and they too had children, and they too had daughters. And the stories of the World War II–era Boston Navy Yard were passed down. They spoke of fighting, and they spoke of loving, and they spoke of acetylene torches, and they spoke of rabbits. Some of the descendants of the crew pursued careers in the Navy.

In 1971, after years of combatting racism and sexism, a Black engineer, Raye Montague, became the first person to fully design a naval ship on a computer. Until then, Montague was employed by the Navy as a typist and was forbidden from touching the office computer. On a day when all the male engineers called in sick, that ban was finally lifted, and she designed a ship on that computer in eighteen hours and twenty-six minutes, a task that typically took her male colleagues up to two years to complete on paper. "I had revolutionized the design process for all naval ships and submarines," Montague said. She became the first female program manager of ships in the history of the Navy. She ran a staff of 250 people. As a result of her brilliance, the process of naval ship design forever changed, and her methods are still being used today. "One of my teachers once said, 'Raye, aim for the stars. At the very worst, you'll land on the moon,'" recalled Montague. "In spite of the system, I was able to accomplish. Not because of the system."

Recently, the naval system launched "strategic evaluations" of the "continued expansion of enlisted female" officers in the submarine force—subs designed by women and built by women are now, more than ever, being inhabited by women. In 2019, seventy-four years after the female shipbuilding crew was downsized, the Navy saw a record number of women requesting to be assigned to submarines—so many that the existing sub fleet can't come close to accommodating them all.

In reference to the building of new subs and refurbishing existing ones to meet the demand, the master chief of the submarine force says, "We have to go faster, because we're leaving talent on the table."

Lieutenant Sabrina Reyes-Dods, coordinator of the Women in Submarines (WIS) task force, sees a future wherein female sub officers will outnumber their male counterparts. "Women make up 57 percent of all degree-seeking college students and earn half of all science and engineering-based bachelor degrees," she says. "Female submariners have always wanted to be treated as submariners, not 'female submariners.'"

## 14

According to biologists, the chemicals oozing within our brains when experiencing the delicious fright of falling, or diving to depth, are the same chemicals that are aroused when we fall in love—noradrenaline, norepinephrine, dopamine, and phenylethylamine. The chemicals collide and twine, braiding feelings of alertness and pleasure with fight-or-flight. The levels of serotonin—that mood regulator—drop dramatically, and we are therefore frenzied and jubilant, ecstatic and intoxicated in a more persistent, unbalanced way. In an "unregulated" way. Thus we associate with the feelings of falling in love the same language we lend to mental illness—we are *crazy* in love, *madly* in love, *insanely* in love.

To sink into deep water is to be able to slowly—if not safely—fall. It is to be the obsessive diver who, enchanted at depth, loses track of time and runs out of air. It is to be the submersible plopped into the sea, its engines revving. The thrill is in the falling, the plunge. We seek—in the roller coaster's drop, and the waterslide, in the landing airplane and in driving recklessly down a hilly road—the arousal of those abdominal butterflies; the thrill, the thrill, the thrill of the descent through space, the feeling of excitement that chemically mimics the disorienting feeling of falling in love.

If we are head over heels, we are upside down, and unstable, and ideally positioned—like an arrow—to keep sinking. The result of this chemical reaction is, according to Dr. Samuel Low, "an addictive rush," the euphoria of which, when absent, we often obsessively seek, stalk, corner. We are "desperate," Low says, "to get more and more of that high." According to writer and Harvard microbiologist Katherine Wu, psychologists have found that the reduction in serotonin we experience when falling in love matches the same kinds of low serotonin levels found more commonly in people who suffer from certain mental illnesses—from OCD to conditions of "overpowering infatuation," to the loss of impulse control and "narrowing of mental focus" often found in diagnosed sociopaths and psychopaths.

Our desperation, infatuation, and impulses—often unfortunately—matter. They affect the bodies of others. How long after some of us are "overpowered" by our chemicals do they seek to regain control by overpowering another? Due to the similar cocktail of chemicals ushered forth during the process of falling in love and the process of indulging in aggressive behavior, the "pleasure of falling in love" can become confused with an attraction to violence via chemical, social, and linguistic influences.

Language, too, matters, and converses with our physiology and our desires. We don't rise in love. We don't move toward love across a horizontal plane. We don't stroll along the sidewalk to meet it, because it's not on our level. And we don't fall *to* or *on* love. We fall *in* it. Linguists and psychologists highlight that we employ this language to illustrate our lack of complicity in the act, and a shirking of responsibility for our responsive actions to it. In this way, "falling in love" resembles the "temporary insanity" defense; the surprised shrugging of someone who is bewildered that they have committed some terrible violence. We enter into a state beyond our control. How long until we justify such violence as our simple innate desire to arouse the feeling of falling in love? Do we just want to feel like we're falling in love? We just want to fall in love. God help anyone within our reach.

≈

Returning to Madsen's workshop, as a practiced journalist having trained her brain to contextualize all that was to come as *story*, to not overlook any detail or consider any offhand moment to be frivolous, Kim Wall curled her fingertips inward until they touched her palm, and like this, she knocked again on the corrugated iron. Inside, Madsen's heart leapt, and he moved toward the door, lifted his own hand through the gaping overlit space to open it.

We don't know what questions Kim asked Madsen as they walked from his workshop to the dock where *Nautilus* was moored. Or how he answered them. Or how she was feeling, or how he was feeling. But we do know—because there are pictures—that the sky beneath which they walked was electric dusky blue, the blue that can appear to be strangely brighter than a sunlit sky at noon. We know that the clouds were thin horizontal lines scarring the sky. We know that, on the horizon, the smokestacks and apartment buildings seemed such tiny black shadows against the expanse of the sky, outlined in the rosy burn of the newly set sun. We know that the water was gray, and then darker gray.

Probably back at the workshop, Madsen asked Kim to step into the orange jumpsuit, the requisite uniform for setting sail on an eccentric's DIY sub. He was already wearing his own faded khaki jumpsuit, which, in the waning light, appeared green, or gray, or the color of the strait. We don't know how either felt when they climbed aboard, but we can guess, differently excited. Maybe Madsen had cleaned up his messes— tidied away the banana peels and apple cores, washed the dishes, made the beds, hung up his clothes. Or maybe the place was still a pigsty. We can guess that their shoes made dull echoes on the steel rungs, the steel floor.

Back at the farewell party, Ole's cell phone vibrated, and on its screen was a picture of bone-white windmills in the middle of the gray water. Ole blinked. His phone vibrated a second time, and there was a picture of Kim smiling at *Nautilus*'s steering wheel. Did Madsen take this photo? Did Kim take it of herself?

# 15

I SAT WITH Albrecht Jotten[*] in his one-room cabin's kitchen deep into the woods and off the grid on the outskirts of Homer, Alaska. Albrecht had "officially" dropped out of the workforce nearly two decades ago at age forty, the age at which he vowed to retire if he hadn't yet achieved "financial independence," retreat into the woods, and, as he put it, "work on my art." Littering the woods around the cabin were his artful manifestations in various stages of function and decay—a green fiberglass spire as tall as the tallest Sitka spruce trees, housing a rickety spiral staircase ascending sixteen floors, each floor complete with a platform on which two grown adults could tightly perch themselves and meditate while enjoying the view amid the trees through face-size portholes; the skeletal and smashed remnants of various seagoing vessels ("failed prototypes," Albrecht called them as he tour-guided), massive ark-size catamarans painted with tie-dye whorls, and homages to Van Gogh that looked no more seaworthy than shipping containers, and a series of increasingly larger and more spherical submersibles.

He made good on his vow, refusing to work on anything but his drive to sink aboard his own vessel into the frigid waters of Kachemak Bay, leaving his beleaguered wife, Nina, to be the couple's sole financial provider. She worked days as a schoolteacher in Homer's elementary school, and evenings as a waitress in a local bar and grill. Albrecht also expected her to tend to the house while he tinkered with his important projects in the woods.

---

[*] Not his real name.

Albrecht was brushstroke skinny with fingers like lockpicks and a patchy sandy beard. His skin was easily given to reddening, and he appeared to have mange. He was mean and short-tempered. Though he had settled in Alaska some thirty years prior, and grew up in the cold climes of northern Germany, he had the mahogany tan and thin, tattered pants of someone who had long ago soured into the malign ease of island living. He was often barefoot, his pants typically cuffed below his knees. His calves were sinewy and one could see their piano strings working even when he was sitting still—usually cross-legged and pontificating about the importance of his beautiful machines. Nina often sat quiet at the small dining table in the center of the one-room cabin and let Albrecht run his mouth, smoothing her blue bathrobe, withdrawing into her exhaustion, the puffy half-moons beneath her eyes visibly darkening with each swaggery syllable uttered by her husband.

"If I drown, then I drown," Albrecht told me late one Sunday night at that dining table as we drank a bottle of tequila, and Nina sighed and looked at her Swatch. Referencing his latest submersible's scheduled maiden voyage (which he variously claimed was only a month, or only two months, or definitely less than a year away), he continued, "They put up statues of people like me when we die, and we usually die early. Visionaries see death, and they chase it, because that's art and that's progress. But I probably won't drown because my sub is perfect. Did my tower collapse like they said?" And here, he pointed a sharp Francis Bacon pinky at his wife: "Like *she* said?"

On the shelves behind Albrecht stood a wood carving of a pelican onto which real pelican feathers had been pasted, and the open jaws of a shark, painted baby blue, into which roses had been long ago placed to dry. "He's an artist," Nina said.

Albrecht said to her, "I'll take you with me."

She didn't smile when she said, "I'll wait for the second drowning."

Albrecht came up with his latest submersible design while sitting at the pinnacle of his fiberglass tower, "talking to the eagles. They think I am one of them now," he said. The tequila was half-drunk, we were three-fifths drunk, and Nina had to get up early to teach Homer's schoolchildren about the overlaps and disconnects between the world's triangles. As the night progressed, the one-room cabin seemed to grow smaller and smaller, squeeze us in like some booby trap.

Albrecht got in my face, and I smelled the ferment on his breath. The tequila was masking something uglier. He told me he was a person who didn't live a life "of desperate quietness," but who had "exploits."

"You watch," he told me, gesturing to a shelf of well-worn Hemingway paperbacks, "my name will appear in books. After the launch, when they see the submarine works, you watch. The Navy will contact me. They'll have me conduct sweeps, ballistic missile research, radar cross-sectioning, flying infrared nuclear signature evaluation, build their anechoic chambers . . . You watch. There's this top secret underwater site in the Behm Canal north of Ketchikan. I know about them, and they know about me. They know I know underwater stealth tech. They know I know how to stay invisible, until I don't . . ."

I sipped my tequila and listened to Nina sigh. Her daughter, also named Nina (whom I had spoken with weeks prior, and who had moved out of the cabin and away from Homer), told me that when she was growing up, her parents enforced a doctrine of "living naturally," which meant being naked as often as possible, shunning the burdens of the capitalist world, which included, apparently, pants and shirts. She told me that the family was always nude around each other when inside the one-room cabin, and she only began to realize some of the awkwardness of this when she and her brother (who was two years her senior) reached their mid-teens. Albrecht soon allowed the enforced nudity—at Nina Sr.'s urging—to morph into encouraged nudity, which became a focal point of the family's increasingly intense arguing. I imagined Albrecht red-faced and naked, his chest sunken and his ribs protruding, his stubble roughly shaved and his mouth angry as he stood nude over Nina Jr. and her brother, slandering their offensive turtlenecks and blue jeans, his fists and penis shaking in the claustrophobic cabin air.

Albrecht poured us a fresh round and told me he was a person who had a "life story worth telling," and who actually wrote it, then bound it with a black plastic spiral between clear plastic covers. As if having stashed it there before my arrival, he pulled his homemade memoir from a box on the floor next to his chair. He told me to take my time with it, and he'd appreciate any feedback on the writing itself. It was 747 pages. I flipped to the preface, titled "Greatest Hits," and perused the first few bullet points.

Albrecht came to New York from Stuttgart, Germany, when he was seventeen. He had no money, so he broke into parked cars and slept in the backseats. He was so attractive, but so vulnerable (in his words) that he was often able to escape prison time by performing sexual favors for the police—"the women *and* the men!" he wrote, "of all the races! They all wanted something from me only I could give, but they could never take my spirit!" He moved to Alaska a few years later to work on the crab boats. He met Nina Sr. in a bar, "back when she was so beautiful, all the men wanted her, but stars find other stars and together they burn bright as bright as the sun, which is also a star. They share their heat with others who can take it. Anyone else would get burned and blind!"

I stopped there and closed the binder and said, "I look forward to the read." On the cabin walls hung diaphanous spills of reindeer skins, framed and crooked pictures of boats afloat at sunset and in moonlight, family pictures in which only the children were smiling. Nina Sr. ran her fingers over the cigarette burns in the cuffs of her baby-blue bathrobe, as if by counting them she was able to divine the late hour. She yawned. "I think it's time for bed," she said. She yawned again. I spotted one family photograph, propped against a set of worn German encyclopedias on the bookshelf, in which Nina Sr. wore this robe, when it was less burnt. Albrecht burped and suggested we throw on our boots and coats and walk outside among his constructions.

On our way out the door, Nina Sr. poured herself a heavy shot of tequila into a coffee mug with a red fighter jet on it. The mist hung thick and cottony in the woods, and the air was cold. All the ghosts were out. The fog twined the top of Albrecht's spire, draped his catamarans, buoyed his latest submersible. So quickly, our hair and our sleeves became damp. Even in this mist, the moon asserted itself like a flashlight through a bedsheet.

"Touch it," Albrecht instructed when we came upon his sub, and I did, and it felt hollow and mean and inconsistently textured—smooth here, scarred there, like a china plate, like a honeycomb, like stretch marks. Albrecht wore a camouflage green peacoat two sizes too big for him, and his body disappeared into it. He looked like an angry wizard. He lit two cigarettes and passed one to me.

"I know you want to see it," he said, and retrieved a ladder from against the trunk of a Sitka spruce. To get to the submersible's hatch, one had to climb to the ladder's top step. Following Albrecht up, I cinched the cigarette between my lips so I could use both hands to balance. The climb seemed to take too long, as if we ascended through something thicker than air—some kind of brine. Albrecht wheezed as he unscrewed the lock and heaved the hatch upward. The hatch groaned as if ancient—some creature that had been slumbering at the seafloor since Verne's time and wasn't yet ready to wake into our present. "Repurposed wood and portholes from local shipwrecks," Albrecht said, citing the materials, "I refired, melted, and re-bent it all myself."

"Clean lines," I said, but I had no idea where those words came from or what I meant by them. Albrecht threw a leg up and over and into the sub, and he went down as if into the water itself—the belly of the beast. I put out my cigarette on the sub's flank and followed him inside, stepping from the exterior ladder to an interior one welded to the curved inner wall.

"It's like we're in a vacuum," I told Albrecht as he turned on a battery-operated lantern.

"Exactly," he said. "There's no other space and time in here besides the sub's."

Albrecht swept the lantern around, and in the erratic beam, I saw a small bar—more tequila—and a control panel cribbed together from various others: Cessnas, automobile dashboards, video game consoles, TV remotes, lawn mower rip cords. A giant repurposed ship's wheel—far too big for this vessel—presided over the buttons and levers. Duct tape ringed the wheel's handles. Acrylic paintings of big-eyed, long-necked girls hung crooked on the wall, all of their hair similarly disheveled, unwashed-looking, and salty. They were all clearly painted by the same artist, and as Albrecht allowed the lantern light to dance over them, I could see, by the signature in the bottom right corner, that the artist was Albrecht himself.

"I sell some of them through Ptarmigan Arts in the village," Albrecht said of the paintings, "but these I keep for myself."

He set the lantern down on the galley counter—another hunk of repurposed wood—and poured us a couple shots. He swallowed his and poured a second before handing me my first.

"They're the saddest," he said, "so I love them the most." In here, the sound of every word we spoke was so quickly swallowed into the vacuum of the space itself that I questioned whether we ever spoke them at all, or if I imagined the conversation. Our sounds were fleeting, small, easily wiped away, the silence and blankness that preceded us so much more powerful and eager to be restored.

"Underwater stealth tech," Albrecht muttered to himself, his eyes staring into his lantern. Then he looked up and seemed surprised to see me, surprised he wasn't alone in this space, as he was accustomed.

"So, what's with this top secret base you were talking about inside?" I asked.

Albrecht smiled. "It has been described as a sinister black silhouette. Only the Navy's Southeast Alaska Acoustic Measurement Facility knows about it. Well, them and the locals. It's hard to hide an underwater base from a population that makes its living off of the sea, you know. But we know how to keep things quiet up here."

A little digging later revealed that the base to which Albrecht referred was indeed housed on a giant black submarine known as SEAFAC, the not quite orderly acronym for Southeast Alaska Acoustic Measurement Facility. The base is essentially a multibillion-dollar American nuclear submarine upon which "new technologies and equipment configurations are tested," though any specifics pertaining to said technologies and equipment, as well as to the nature of the testing processes, have been scrubbed.

According to Tyler Rogoway, a journalist and historian of military strategy, technology, and defense (especially as pertaining to issues of aerial and underwater procedures), in the article "The U.S. Navy Has a Critically Important Submarine Test Base Tucked Away in Alaska," SEAFAC's primary mission is to measure the "acoustic signatures" of nuclear submarines. An acoustic signature, to simplify, is the collective noise something makes, and the effects of that noise on both the thing itself and the surrounding environment.

Humans, for instance, are blessed with the stapedius muscle, a nugget of gristle less than one millimeter long, hidden in the depths of the middle ear. It is the smallest muscle in the human body and is responsible for mitigating our own various acoustic signatures, basically dampening the amplification of our own innate sounds, which, if taken together undampened, would be so overwhelming that the noise would drive us mad.

The stapedius is the tiniest thing inside of us that bears the greatest responsibility for our sanity, and works alongside the brain to modulate and control our impulses and our ability to civilly function without being subsumed by our body's sounds, our inherent biological loudness, the screaming voices in our heads and hearts and bellies.

"Acoustic signature," Rogoway writes, "is the primary survivability factor of the modern submarine. Every aspect and component of a submarine's construction takes this reality into account. And it's not just about quieter mechanical systems, pump jets, or coatings on a submarine's hull, but also about how systems are mounted inside a submarine to isolate vibrations and sound waves as well. Even the shape of a submarine and the water flowing around it can be a noise factor."

"I love also how quiet it is in here," Albrecht said, staring into his lantern. "It's so quiet, every little noise can be loud."

Here in the sub, though we weren't at depth, we certainly didn't seem to be earthbound either, amid the trees, a stone's throw from the cabin. We didn't seem to be on the same plane on which restaurants and libraries and schools and banks existed. We seemed, instead, insulated from the machinations of human life on land, from mundane commerce, bland economic and social transactions; insulated from a pressure toward politeness. Less ashamed of the thunderbolts of our ids. Less apt to disguise their loudness, and subsequently, more willing to curdle the ears of another.

I heard my heartbeat, and I was convinced it was beating out of control. I tried to calm myself down with thoughts of things that seemed an antidote to depth and sinking and claustrophobia and claustrophilia—images of open sky and soaring birds. I thought about the Nicobar pigeon, the closest living relative to the dodo bird. It resides on far-flung islands and faces no natural predators. Having no need to conceal itself, it developed brilliant, iridescent feathers. I thought about how predation can squash exquisiteness. I wondered if isolation is a price worth paying for beauty. I wondered what immersion into the sea has to do with any of this.

"Who's your favorite genius?" Albrecht suddenly said, his saliva crackling like electricity.

"Um . . . ," I said.

"It's okay," Albrecht said. "It's a hard question if you haven't thought about it. You can take some time. Me? I've been thinking about this a lot."

Albrecht's voice in this space was loud and sharp, and shortened, as if a firework exploding without any residual reverberation. Each syllable ignited. Then deafened. Then disappeared.

"Your last name . . . ," Albrecht continued. "Are you a Jew?"

My stomach dropped, as it tends to do when someone asks me this question. I never know what's coming.

"I am," I said, swallowing hard.

"Oh, well, then maybe I shouldn't say this, but you're a writer, so I think you'll be able to understand," Albrecht said. "For me . . . and I know it's not popular to say this, but it's true . . . the biggest genius is Hitler."

I didn't know what my body was doing, what my hands and feet were doing, but I remember my face getting very, very hot. There I was, trapped in this dark greenish nightmare of a submersible, surrounded by crooked trees, and Albrecht was so close to me I could have touched him without fully extending my arms. I could have hit him, or pulled his hair, dragged him to the metal floor and strangled him to death, picturing, right there in that tight balloon of a space, the faces of my ancestors murdered in the concentration camps.

"I mean, you're a smart person," Albrecht said, "so you have to admit it. You may not like the things he did . . . but look at what he did. Look at everything he was able to do, and all of the people he influenced. It was such a . . . a . . . careful plan, you know. Everything had to go right. Everything had to be perfect. And he did it. Everything went right. It was perfect. *Perfect.* It is the example of the best genius of our times. I have to admire it."

## 16

THE STEEL-GRAY WATER bubbled around *Nautilus*'s hull. How good was Peter Madsen's autopilot system? Good enough for him to abandon *Nautilus*'s controls soon after detaching from the dock and pulling into the strait, so he and Kim Wall could climb back up, open the hatch to the dusk and the breeze, and poke their upper bodies into the air as they were sailing past the spot where Kim knew Ole and her friends would be waiting and watching.

And now there they were onshore, whooping and clapping and hoisting their plastic cups. *Nautilus*'s conning tower from which Kim and Madsen emerged was, like most conning towers, cylindrically shaped, and in this light, their bodies appeared to Ole and friends as if two candles cresting the top layer of a wedding cake. Facing the sub from the shore, Madsen was to the left of Kim. Their bodies, from their sternums down, were inside the body of the tower, but they both lifted their arms outside into the air, bent them, and rested their elbows on the tower's flat surface, two chickens mid-flap. The *UC3* logo, rendered in white blocky font and shadowed with gray stenciling, allowing for the illusion of depth, was painted onto the tower's left side, and Madsen lorded over it, the letters and number each about the height of his own body from the waist up. The wind at this point over the strait couldn't have been too strong, as the Danish flag to Madsen's left dangled limp on its pole. He stood between this flag and Kim, each about one foot from his bent elbows. To Kim's right were the skinny twin columns of the periscope and the radio antenna. And she stood between them and Madsen, each about one foot from her own bent elbows.

The symmetry is shocking, or staged, or forced by what we don't see in the pictures—the tight parameters of the conning tower's cavity, and the ways in which Kim and Madsen had to contour their legs on the ladder inside its gullet. They stood so close to one another. In the picture, they are still standing. Their feet must have shared space on the same rung, the vibration of the sub's engine shuddering their shoes in exactly the same way, compelling the bottoms of their feet to similarly itch. Her white shoes, his brown ones. They took in air and let it out. Their hearts beat. Madsen's exhales were louder than Kim's, more voluminous. His thoughts were in a deeper, darker place.

Kim waved wildly at Ole and her friends, and maybe, though the light was minimal and her intimates were silhouettes, she could distinguish who was who. She waved and she waved. She was in love, and she was pursuing her story. She was moving to Beijing. Beneath Kim and Madsen, in the body of the sub, also shuddering with the vibrations of the engine and swaying with the sea, newly stocked, were the orange-handled handsaw, lead pipes, straps, zip ties, a knife, and several sharpened screwdrivers. Their handles rattled and whined. The coffee Madsen had prepared for the occasion sloshed against the glass walls of the pot. The chocolate chip and oatmeal raisin cookies he had fanned out on two paper plates vibrated—as yet unbitten—against the metal counter.

The previous night—August 9—Madsen had conducted internet searches for *executions* and *dismemberment; beheading, girl,* and *agony.* And earlier that day, on August 10, he had invited, via text message, one of his few female interns to join him that evening on *Nautilus* for a date. Initially she accepted, but as the hours progressed, she thought about how Madsen had seemed even more erratic than usual these past few days. Part of his invitation—which she thought at first was a joke—read, "I have a murder plan ready which is a great pleasure." She decided to text him to cancel. When she did, Madsen decided to contact Kim.

The gray water of the Øresund Strait pulled taut like a sheet, and the distance between Kim on the sub and Ole on the shore seemed so easily crossable, even if one wasn't the strongest of swimmers. She was right there. In the picture Ole took from shore, Kim has dropped her arms after waving, her elbows having returned to the tower's surface. Her smile, though, persists. Her shoes are rattling. Her nose is cold.

And she's looking directly at the camera, having somehow perfectly ferreted out the location of the lens from the shadows. "In stepping onto that submarine," the journalist May Jeong would later write, "Kim was doing what any reporter onto a good story would have done." Right there. "I'm still wishing for a lesser tragedy," Jeong would write. "I wish for a different story."

Madsen can't bear to look. Maybe he had been fantasizing about this moment for decades. How does Madsen experience nervousness? Shame? How did he sleep the previous night, and how did he sleep on this night? Did he dream of flying or sinking, or puttering around in his workshop making Nazi jokes; of the birds and the bones of the Tissø shore; of the sour breath of his father? In the picture, standing only one foot from Kim, he is turned away from both her and her friends onshore, only his left ear visible, the ridge of his brow, the curve of his chin, the ruffled reddish hair on the back of his head. He is staring past the Danish flag out to sea, at the distant horizon, across the measurable silences and depths they still had to cross, his heart inside of him doing whatever it was doing, quietly for now. This was the last time anyone besides Madsen saw Kim Wall alive. And Madsen couldn't bear to look. In the photo, Madsen still can't bear to look. In Kim's hands (and it's tough to make out) is what appears to be a piece of paper, carefully folded into the shape of a long-beaked bird.

≈

The sky over Refshaleøen grew dark, and the goddess constellations made themselves known. The stars twinkled and shot and lived and died. Kim and Ole's guests were still partying quayside and were growing progressively tipsier. Ole's phone intermittently buzzed with text messages. "I'm still alive btw," Kim texted Ole. "But I'm going down now. I love you! He brought coffee and cookies tho." Madsen told Kim to help herself to the coffee and cookies. As she did so, Madsen pulled his cell phone from his pocket and sent a text to his friend and frequent rocket project collaborator, Christoffer Meyer—on whom Madsen had once bestowed the coveted title of Rocket Madsen Spacelab flight director—canceling the plans they had made to take an afternoon trip on the submarine the next day.

Meyer, a sturdy, full-faced man with a worried forehead who favored blue-and-white houndstooth button-down shirts with stiff collars, thought nothing of the cancellation, as Madsen was prone to such fickleness. "I had no alarm bells ringing," Meyer later said, his voice breaking. "Peter definitely has a temperament. But I have to be fair to Peter and say that over the years he has become much better at controlling it. I've seen him shout and scream, but the fact is, I've never seen him lift a tool or clench a fist or grab someone and say, 'If you don't stop, you'll get a pair on the box.' I've never seen that."

Seventeen days before Madsen and Wall set sail on *Nautilus*, Meyer threw a surprise birthday party for his wife in the backyard of their home in Farum, a forty-minute drive from Refshaleøen. Madsen attended, and the partygoers mingled on the manicured lawn, ate cake, and drank beer beneath the white outdoor tent Meyer had rented for the occasion. "Peter could well be the center of the party," Meyer said, "but this evening, he did not feel very comfortable because his cat was sick. If the cat was sick and ill, [Peter himself] would not feel very well. Then he actually just wanted to be allowed to take care of it. And I know [that at the birthday party] he asked others for money because he could not afford to pay for the vet."

On *Nautilus*, Kim texted Ole a few more pictures: selfies in the hull, the puke-green paint, the dirty counters, a plate of cookies, the unkempt bunks, Madsen in shadow.

At about 10:00 p.m., when the messages stopped, Ole wondered how long this submarine ride was going to take. They had to get up early the next morning, but he knew that Kim's dedication to the story would compel her to sacrifice sleep. She had been waiting so long for this interview, and it must have been a fascinating one. One of their friends suggested packing up the grill, doing a cursory cleanup of the picnic area, and moving the celebration to a nearby bar—the Halvandet, a trendy harborside hangout with outdoor seating and live DJ (replete with silver and purple strobe lights), where the young intoxicated crowd often plays beach volleyball, Ping-Pong, and archery among the silhouettes of creaky boats and the detritus of defunct industry. At Halvandet, the group drank craft beer from clear plastic cups and craft cocktails from little glasses and danced and had to lean in close to one another's ears and shout if they wanted to be heard.

Ole, the DJ's lights reflecting from his cell phone, kept checking the time. He was getting anxious. Kim should have contacted him by now. She was so good about keeping in touch. Some guests, a little too drunk, walked to the water and heaved, before plopping down at an open spot on a bench at one of the turquoise picnic tables, wiping their mouths and staring at the strobe lights flickering from their shoes. Ole paid them little attention and kept one hand in his pocket, on his phone, waiting for it to vibrate. The music was beginning to give him a headache. People were laughing, sipping, frolicking in the sand.

Without saying goodbye to his friends, Ole left the bar and walked the alleyways, hemmed in by the tall green iron fences of the dead shipyards, the barbed wire drooping like spiders' webs at their tops. He walked past the nighttime parking lot, where the red-and-green hop-on/hop-off Stromma sightseeing buses lay dormant, beached and boxy, petrified as if some unfortunate pod of whales; past the graveyard for old construction signage, road barriers, reflective orange cones having gone limp, half-melted by sun and road tar to the point that they appeared to Ole, in the dark and in the sea mist, as stalagmites.

When he arrived at their apartment, the windows were dark, the door locked. He fished into his pocket for his keys, went inside. In the small kitchen, dirty dishes, slick with dried ketchup and congealed grease, were stacked on the counter, and wineglasses, stained with droplets of red wine, lipstick, and fingerprints, filled the sink. Ole washed a few, decided to leave the rest for the morning. He checked his phone. He went into the bedroom, got undressed, considered the rumpled sheets, the pillow on Kim's side of the bed. He checked his phone. He got into bed, turned off the nightstand lamp. In the dark, he saw things, car headlights playing on the wall, shadows rising and falling. He waited, listened for the click of Kim's key in the lock. Listened for the vibrations of his phone. The building hummed and fell silent, hummed and fell silent. He closed his eyes, but he could not sleep.

He sat up in bed, inhaled. He wiped the sweat from his forehead and put on his clothes. He looked at one of Kim's texts from earlier that evening, around 7:30—a picture she took from *Nautilus*, of windmills on the water. In it, the sea dominates the image, and the windmills—sandwiched between the gray sky and the darker gray water—look overwhelmed and feeble and too skinny, insubstantial as the legs of a mosquito. He paced the bedroom, talking to himself, working through scenarios. He should have gone with Kim, he told himself, should have agreed when she invited him to join her on the interview, ride the sub. He was "insanely close to saying yes," he would later say, but felt compelled to stay back with their guests. He suddenly thought of how he and Kim kissed each other before her walk to Madsen's workshop—how it was stronger and deeper and lasted longer than their typical kisses, because, as Ole would later speculate, "she was going out to sea," not simply "out for, say, ice or lemons."

But there, pacing the apartment, staring at the lit-up image of the windmills until his eyes watered, Ole thought of Kim, and the sea, and soon, before he realized what he was doing, he found himself outside the apartment building in the middle of the night, unhitching his bicycle from the rack. He pedaled the streets of Refshaleøen breathlessly, his speed driven by his anxiety. He pedaled on pavement and gravel and dirt. He pedaled past rows and rows of converted shipping containers. He called out Kim's name in a thin voice, over and over. He heard no answer but the howl of the wind, the roar of the sea.

He pedaled to the seaside, and along the quay, past the spot where he and their friends ate and laughed together earlier that day, now empty and dark, and scored only with the sounds of the boats moaning against the docks, their tall masts ghostly against the sky. He biked along what seemed to be every possible path Kim could have taken. Then he biked along them again. Desperate, at approximately 1:45 a.m., he stopped his bike, leaned it against a chain-link fence, and called the police. Morose and dashed, his damp shirt lifting from his body in the wind, Ole Stobbe rode his bicycle back to the apartment to wait for them. In the kitchen, he clutched his cell phone. His grip tightened. The screen grew foggy with the heat of his hand. Two and a half hours later, at 5:31 a.m., he would use that phone to call Kim's parents. Her mother would wake and pick up. Her father would sleep through the ringing.

# PART TWO

## 17

Through the traffic outside Oakland, California, behind silver mirrored sunglasses, Shanee Stopnitzky—one of the few "non-dudes" in the p-sub community—drags *Fangtooth*. The small white home-built submersible shimmies in the trailer behind her white pickup truck as it rockets past medians decorated with barbed wire and bright bubble-letter graffiti. PEACE LOVE SURF, reads one section of concrete. FUCK THE QUEEN, reads another. The yellow utility strap holding *Fangtooth* in place vibrates in the wind. The sub's pink-painted propeller slowly turns. Stopnitzky has to shout to be heard.

"You get all kinds of reactions," she says about her sub. "I'm concerned about causing an accident because people film us a lot while they're driving. It takes a moment for it to hit them—that's an actual submarine." Stopnitzky laughs, her smile stretching across her face. "She's also just really cute and fun," she says about *Fangtooth*. She feels the sub is "more psychologically accessible because of its playfulness."

Stopnitzky laughs easily and deploys words like *wonder* and *awe* with such aplomb that a listener can't help but feel such emotions well up within them as she speaks. Her hair is black and shoulder-length and feathery, bangs short and squared off. If this were high school, she would be the coolest and most self-possessed of the goth nerds, a punk Linda Ronstadt. After five minutes with Stopnitzky, the world feels larger, mysterious, amazing. She is a person one doesn't want to stop being around, for fear the world will become matte in her absence.

*Fangtooth* used to be painted yellow, but was repainted white because, Stopnitzky says, "We never want to hear the Beatles song again." The white will serve as "the base coat for her new paint job where she's gonna get painted by a street artist and will be a much more accessible little creature than she was before with her Captain America shield," which formerly graced the outer part of the hatch.

*Fangtooth* groans on the turns, as if the imposing jaw of the deep-sea fish after which it is named were creaking open. The fangtooth, which has no known relatives, has, proportionally speaking, the largest teeth of any fish in the ocean. They are so large and so sharp, the fish can never close its mouth, and their sharpness is always on display, the fishes' jaws appearing haggard and forbidding, their teeth serrated and roped with mucus, capable of impaling fish and squid larger than themselves on their faces. Even their scales are more like sharp plates. The fangtooth has had to evolve to adapt a pair of opposing sockets on either side of their brains in order to accommodate their longest two bottom fangs. They rove the gloomy depths by day and rise only at night to feed by starlight. Before evidence of their existence was documented, the fangtooth—which is among the deepest diving of all fishes, having been found to thrive at depths of sixteen thousand feet—belonged to the realm of mythology as a pet of the devil. Still today, we know so little about the intricacies of their behavior. Still today, the adjective often accompanying its description is *monstrous.*

*Fangtooth* is classified as an "experimental class sub," which means it isn't beholden to the costly certification process that can make a commercial-class sub, for instance, cost many millions of dollars. "In the United States," Stopnitzky says, "I would estimate there are maybe thirty experimental-class subs. It's hard to gauge because a lot of people do it without telling other people."

She drives past a scrapyard, and a decaying school bus, and a long line of auto parts stores and collision garages. Stopnitzky is wearing a black Community Submersibles Project T-shirt, the name of the company she founded in 2018 and directs—a GoFundMe-supported "submarine collective" through which "anyone can use the submarines," including the two-person *Fangtooth,* and another christened *Noctiluca,* after the marine species of dinoflagellate. *Noctiluca scintillans,* a unicellular microorganism, can bioluminesce, exuding an unearthly red or green glow. When the species blooms at night, an entire body of water can appear to be lit from within. But the species is also a siren. When it blooms and glows so beautifully, it is also overtaking the body of water in which it lives, eliciting such environmental hazards as toxic red tides and eutrophication, which can pollute a water body in the same way that industrial wastewater, fertilizer, and sewage can. Stopnitzky purchased *Noctiluca* the sub for $60,000 from a "private person" who had bought it from the anti-whaling activist consortium the Sea Shepherd Conservation Society, which originally painted the vessel black and white to look like an orca "to scare off gray whales to keep them from getting hunted." Apparently the Sea Shepherd folks secured *Noctiluna* from the Swedish Army. The sub has a crush depth of twelve hundred feet, and Stopnitzky hopes "to do some retro fitting to enable her to go deeper . . . to allow her to go *much deeper*."

"[The sub] was a money pit for him [this 'private person']. I mean, he wasn't enjoying it. It was a source of anguish. A friend jokes that to get really great deals on Craigslist, you put 'wife' in the search parameters"—her voice labors to keep the laughter at bay—"because once a wife decides something's gotta go, it's, like, really gotta go, and that's how you get the best deals, and that's what happened. I actually was in communication with his wife, and she was like, *Please, please for the love of God, take this away from us.* We bought [it] in sort of disrepair and we've been repairing . . . building out [the] life support systems. I am totally self-taught." She chuckles and shrugs. "My credentials arrrrrre none. I am in a state of learning for sure . . . It's so funny to me," she says, "you could just buy a submarine . . . ," and she throws her head back and laughs, briefly losing herself.

But then she catches and contradicts herself: "Well, you can't just . . . buy them, usually," she says, now carefully considering her words. "You have to sort of . . . demonstrate your competence so, you know, whoever built the sub originally doesn't wanna have to worry something terrible is gonna happen . . . So you kinda need to be a part of the community. It's not uuuuusually an option to just show up and buy one. I was very persuasive," she says, and laughs again.

On her T-shirt is a submarine sunk below the squiggly line of the water's surface. In the tableau, it's night, the moon and stars glowing at the shirt's collar and shoulders. In the sea are a riot of odd and lovely creatures, bannered by the script GET DOWN WITH FRIENDS. "The coolest thing I've ever made," she says, "is this submersible community."

Stopnitzky is also an "experiential" installation artist. Her artwork is interactive and performative, and she stages events around her pieces, using them to further solidify her submersible community. Repurposing an old pontoon and affixing it with "twenty-four fluttering wings controlled by a swing-set," a trancey sound system, thirty-five hundred multicolored programmed LED lights, a gold glitter anchor, a mosaic floor made out of smashed DVDs, and "fire effects," Stopnitzky has created what she dubs the Artemiid Art Boat. This floating sculpture was available to rent out for parties, the proceeds of which went back into her submersible agendas. At sea, at sunset, the Art Boat resembled the offspring of Pegasus and a stegosaur, the upper back of which was flayed to the spine and ribs. Something not quite of this world. Some gorgeous dream coat of a decomposing cartoon made manifest. A living carousel.

Stopnitzky has fashioned an exhibit using thousands of living butterflies to evoke the "dynamics of chaos," and a "motion-responsive glow-worm chandelier," and *The Firefall:* layered shards of charcoal embers "poured from a balloon-mounted receptacle" to resemble a glowing waterfall. She also makes glowing "pandora" flowers (made with actual flowers—gladioli ideally—a black light, and fluorescent highlighter), algae watercolors (created with a process she calls The Squishening), and "fire bubbles" called Indestructibles. To create the latter, Stopnitzky says, "you bubble propane into soapy water, making propane-filled bubbles. You scoop them into your hand, light them on fire, and they burn up quickly enough to only singe your arm hair but not hurt you. Advanced level is getting a bunch of people in a circle, then get everyone to scoop some bubbles, and then you can pass the fireball from person to person really fast until the last person throws their fireball back into the cauldron and all the leftover bubbles ignite into a massive and rapid fireball.

"[I] try to be a builder and propagator of objects and sentiments whose only purpose is beauty, whimsy, or silliness," she says.

She weaves the pickup truck through traffic like a dance, *Fangtooth* trailing like the train of a wedding gown. "We had cops do that actually one time," and it takes me a minute to realize that she means: gawk at the sub. "We thought they were gonna pull us over, and they were actually, like, driving really slowly, filming us."

She peers over her shoulder to change lanes, and she too can't help but gawk at the sub. "Submersibles are not just for war and science," Stopnitzky says (though she is a complex systems marine scientist). "We want to emphasize wonder and encourage people to seek out and value awe as a fundamental human experience, using submersibles." She also describes herself as a "mother of submarines, experimental artist, and evangelist of embodying wonder" whose mission with the Community Submersibles Project is to make "deep-sea exploration by submersible available to the not-rich wonderers of the world."

Stopnitzky got serious about possessing her own sub after a diving trip to Honduras. "I think curiosity makes everything better," she says. "I've always been obsessed with the deep sea and I've always wanted to have this experience of going in a sub," which she was finally able to do off the isle of Roatán. She dove down to two thousand feet, to the bottom of an ancient coral reef. One can tell: even while driving on the California freeway, Stopnitzky is still down there in her head, amid all that buoyant silence, her breath fogging up the lens of the submersible's porthole. She changes lanes as if on autopilot. In the car she passes, a couple of kids in the backseat gawk at *Fangtooth,* their own breath fogging up their own window glass. Stopnitzky's voice lowers in pitch and a seriousness crosses her face, and I feel as if I may weep when she says, "It was one of the most profound experiences of my life," her tone itself invoking a profundity and gravity, like a chasm opening on the seafloor, through which thousands of fish emerge, waving their translucent orange fins. "Hands down, I am completely obsessed," she says. "It's all I wanna do with my life."

In 2018, she left her PhD program in marine biology at the University of California, Santa Cruz, in order to work on her own "experimental submarines" and to address issues of submersible accessibility.

"I'd been unhappy in the program for a long time," she says, "and I'm actually obsessed with research. I fully expect to still work as a scientist."

The public, of course, can dive only after a rigorous training session, and under the guidance of Stopnitzky and her team of ragtag, sunburnt technicians. "Being in a homemade sub is actually less scary in a lot of ways," Stopnitzky says, "because you're only going to shallow depths, so the risk of any kind of pressure injury is much less." At least within the confines of *Fangtooth*, the chances of sustaining barotrauma—pressure-induced damage to body tissue—is virtually nil.

She drives past the Whole Foods and the Safeway, the La Quinta and the DoubleTree, the Donkey & Goat Winery and Tasting Room, before pulling in to the Berkeley Marina, where people in white linen shirts drink white wine at the adjacent yacht club, and people in sun-hats hold binoculars to their faces at the adjacent Shorebird Park Nature Center, and families in tank tops and flip-flops play volleyball in César Chávez Park. The masts of the sailboats stand like sentries against the sun.

In 2018, Stopnitzky says, laughing, during the last weekend in April, *Fangtooth* disappeared from its space in the Berkeley Marina. "Somebody stole it and took it for a joyride, we believe . . . It was hectic for a few hours as I called every single law enforcement person and asked them if they had—actually, yes, for real—seen a yellow submarine. . . . And they all were like . . . *Are you kidding?* [And I was] like, 'No, no, no, I'm serious. Yellow submarine. Someone could be dying in there right now . . .' I was really frantic 'cause I was scared someone was gonna get hurt. It's super easy to get yourself into trouble if you don't know how to . . . ," and Stopnitzky swallows hard and pauses.

This kind of pause is typical of DIY submersible enthusiasts when they bump up against a moment in their story when they have to confront the real dangers lurking in their enthusiasms in spite of prior assurances of safety. Such pauses are often rife with pathos, lending—via intricate consideration of their own obsessions, and the benefits and consequences thereof—the "pauser" a greater and more complicated sense of humanity.

". . . rescue yourself out of these things," Stopnitzky finishes.

*Fangtooth* was found abandoned some eight miles away in open water near the Bay Bridge, the joyriders having made their unlikely escape. The Alameda County Fire Department pulled the sub ashore and turned it over to Berry Brothers Towing in Oakland (the Google reviews of which include "Prepare to be ripped off"), who held *Fangtooth* hostage.

"We [had] to come up with $2,000 to get it out of the impound lot," Stopnitzky says. With her characteristic enthusiasm, she was able to solicit the funds from a group of supporters, fork over the ransom to the villainous Berry Brothers, and have a happy reunion with *Fangtooth*.

Now, as Stopnitzky untethers *Fangtooth* from the truck and eventually, with the help of a couple of her flannel-clad techs, transfers it to the water, her excitement begins to overflow, her words interrupted by sounds—quiet whoops and cheers and yelps, *ooooo*os and *eeeeee*es and *rrrrr*rs. It's truly a joy to behold such joy. "Oh, oh, oo, ee, oh, rrr, it's so fun," she says. "We're gonna hit the bottom; it's gonna be fun."

≈

The sea bottom beneath the Berkeley Marina, while beautiful and worthy of Stopnitzky's excitement, also harbors the remnants of tragedy: urn shards; metal screws and plates that once held people's bones in place; dentures that resisted the heat of the crematorium; the charred wooden and metal crumbs of a sailboat that mysteriously caught fire in 2022, claiming the life of the man who lived on board; scraps of the swimsuit of the man who drowned here in 2012; bits of the gray whale that died here—also in 2022—the sight of which, at a distance, resembled an overturned rowboat blanketed in ravenous seagulls, but upon closer inspection so alarmed the marina-goers that the carcass had to be towed farther out to sea, so it could decompose without disturbing the residents.

The waters around the Berkeley Marina also see the official burial-at-sea of Oakland's unclaimed dead. On his decrepit yacht *Orca III*, Captain Curtis M. Brown ferries the cremated ashes of the area's "homeless, the mentally ill, the elderly poor, the anonymous." To Oakland's coroners, they are "indigent cases," and approximately two hundred such burials take place per year on average, but the numbers are seeing an uptick, "as is the case whenever there is a downturn in the economy," the *East Bay Times* reports. "Without anybody really noticing," posits an article in SFGate, "San Francisco Bay has become an enormous burial ground."

There are so many things interred in this water. Captain Brown has been jokingly nicknamed Charon, the psychopomp ferryman of Hades who, in Greek mythology, ushered the dead across the river Styx. "It's a small world out there on the water," Brown says. Every three weeks, a delivery truck arrives at the marina, loaded with blue rubber storage bladders containing plastic urns. Oakland's Evergreen Cemetery works in conjunction with a company named—alarmingly, in this context—Nautilus Cremation, which charges the City and County of San Francisco fifteen dollars per cremated person, a cut of which goes to Captain Brown. The owner of Nautilus Cremation, Buck Kamphausen, invented a device he calls The Disseminator, a big metal colander attached to high-pressure water jets. If the cargo is particularly heavy, and the urns numerous, the contraption makes it simpler to disperse a mountain of ashes. Kamphausen, the child of ranchers (who says he is so named because he was, from birth, "a little buckaroo"), also builds cemeteries and mausoleums. "It's a good, solid little business," he says. *Orca III* has a Disseminator set in its stern.

Once the urns are loaded, no matter the weather, *Orca III* sets sail. In a wistful tone reminiscent of Stopnitzky's, one of Brown's crewmates finds the beauty in these trips to sea. "There's something calming about the water," he says. "We've done it at night, in the fog, in the rain. You have to say that if there is something spiritual about it, this is one of the most beautiful places in the world to have it done."

For these ceremonies, the captain dresses respectfully for his duties in neat black slacks and a white collared shirt. If it's sunny, he'll wear his dark wraparound shades. "Over the recent years," he says, "I'm sure there have been more people scattered in the water than buried in the Bay Area . . . We have a moment of silence. Then we place some flowers in the water and the ashes sink."

There are so many stories at the bottom. Shuddering far beneath *Fangtooth* now are ashes, flowers. The boats in the marina rock, *Orca III* creaking like a gallows. Stopnitzky smiles from ear to ear. The sobering stories and remnants notwithstanding, the bottom for her remains a place of richness and astonishment bordering on reverence—a place simultaneously otherworldly and more *of* this world than our reductive terrestrial version. Why *wouldn't* one want their ashes consecrated in such a place? The water is such a lovely color—the glassy green of a soda bottle; some big jade urn.

"So, I say I was a marine scientist from the day I knew that a fish existed, which is about four years old. It's the first time I can remember wanting to be underwater all the time," Stopnitzky says. She will later brandish a photo of herself at that age, sunk into a life jacket way too big for her, only half of her head poking out from the orange pillow. In the picture, she's sitting in a tiny sailboat, a skinny blond boy—her older brother perhaps—behind her, his bare arm resting on the boat's wooden rail. Another blond boy—another older brother—stands on the border where the beach sand meets the water, helping their father, who is already knee-deep at the bow, maneuver the boat into the sea. The photo is blurry, and the father is so skinny, clad only in a short turquoise swimsuit, his bones and muscles jumping from his torso, his sandy hair disheveled. He's bent, reaching toward a white rope, his body casting a shadow over the toddler Stopnitzky, whose gaze remains fixed and forward, as if hypnotized by the open water. "[Today], both of my brothers are scientists," Stopnitzky says. "They copy me."

"I've done about thirty-five hundred dives," she says. "I've now spent a year of my life underwater, in aggregate." Technically, Stopnitzky, who has logged in excess of nine thousand hours at depth, has spent *over* a year of her life underwater. A year of days, a year of nights. A year's worth of dreaming. Seven hundred and thirty cups of morning coffee for the modest imbiber. Five hundred forty-seven and a half sleeping pills. Three million six hundred and fifty thousand steps. The entire life span of the panther chameleon, the *Paedophryne* frog, the mosquitofish, the ruby-throated hummingbird, the common mouse: underwater. Fourteen lifetimes of the housefly, twenty-six dragonfly lifetimes, twelve lifetimes of the worker bee: at depth.

"The actual moments you spend underwater are very emotional and beautiful, and for me, it's . . . it's more psychologically soft . . . As soon as we're submerged, as soon as the hatch is underwater, I transition . . . I've also always had this dream of building an underwater house, and I feel like I now have the skills to do that . . . Ever since I was a little girl I wanted to live in an underwater house. It's been part of my lifetime arc and goal to move underwater."

Stopnitzky is quietly working with a team of engineers to make her underwater human habitat dream a reality. The project, which she refers to as "the ultimate objective," is being fully funded by a group of anonymous donors, and—perhaps because its development is in the early stages—Stopnitzky is reluctant to provide specific details about it, only that the habitat would be available for "long-term underwater stays," and that its tentative name is La Chambre Bleue. A little digging reveals that Stopnitzky, on May 5, 2022, put a posting on MIT's Center for Ocean Engineering job board seeking "part-time ocean engineers for consulting on a project for an underwater habitat for human occupation . . . a type of structure that has never existed before." Potential applicants are asked to submit a "brief (less than a paragraph) introduction to their background and interests." "This is a private venture," the posting stresses, but divulges that the location for the underwater habitat is to be "45' underwater in the tropical Indian Ocean."

"The subs are a stepping-stone to the underwater house, technologically," Stopnitzky says, "and they're also the fulfillment of this dream of exploring the unknown and exploring the underwater realm . . . I kept building bigger and weirder things on the water and finally felt I had to amass the resources in my life to actually build an underwater house . . . It's definitely gonna happen. Things are moving really fast."

I can't help but pry and push and repeat my questions until Stopnitzky reveals more about the project. "It's intended to be an experimental space . . . There will be natural purposes that emerge out of it, but we're specifically keeping it purposeless. There's no specific objective but to be a human and be in that space. And whatever that produces, will be. For me, these are really fundamental human experiences."

Stopnitzky laughs when I ask her about what the thing is going to actually look like, physically. The schematics are still in development, but she wants the space to be big—an underwater mansion; big enough to house twenty-five people for about a week, or five people for a little over a month, "so we can recharge the life support," she says. "It will be at about fifteen meters of depth and embedded into a coral reef." She intends to build the house of steel and a series of interconnected acrylic tunnels, "for maximum visibility." Stopnitzky is scouting out probable areas near the Philippines, Indonesia, and Southeast Asia, and feels that from start to finish, the entire project will take three to five years, at a cost of seven to fifteen million dollars (which would include fees paid to the local government for a permit to lease the undersea space).

"So I stole them [my crew] from the submarine industry." Stopnitzky laughs. "We have five working on it full-time, and we have subcontractors—this team that specializes in space habitats, which is really fun. So, five core people, and probably a wider net of maybe twenty-five at the moment. There's gonna be a lot more as it develops."

For instance, Stopnitzky wants to develop a "complicated array of window wipers," to keep tiny marine creatures from colonizing the panes. She wants to create a facade "to encourage reef growth," so the reef itself can eventually mask the enclosure. While it was originally designed in the shape of a star, Stopnitzky is now conceiving of the space more as "a giant permanent submarine."

Stopnitzky admits she actually *endures* time on the surface, as if it's painful for her—a too bright, irritating, unnatural way station. On the surface, Stopnitzky says she doesn't "feel whole." "I definitely feel separated from my place when I'm on land. So, the natural solution to that," she says, is to live as long as possible underwater.

She mentions, her voice wavering with excitement, how the underwater house would allow her to stay at depth for longer periods than would the submersibles.

"It's really hard when you're down there to want to come back to the surface ... No. I don't want to go back. You know, this is my place."

One wonders in this way to what degree the underwater realm serves the role of convent or monastery, a meditation retreat, a cloister, a version of agoraphobia strangely set in such an expansive space. A place to be a human, away from the rhythms and ornaments and structures of the human world. A place where we shouldn't be, a place our bodies weren't made for, wherein we require machines in order to respire. Though I adore Stopnitzky's vision, and especially her notion of the purposelessness of the underwater house—for her and many others the time spent at depth manifests in the humility and wonder of witness—I've probably watched too many Bond movies and can't help but think also of villainy, and the dark side of chasing "legacy." Could this be another version of misguided human stewardship and manifest destiny, akin perhaps to colonizing the moon? If such a structure begat others, I wonder how *we* would change, evolve to better greet the pressure of the deep.

According to Dr. David Sawatzky, an expert in diving medicine, "narcosis has been a known effect of breathing air under increased pressure," which is how we breathe air at depth. Sawatzky says that "as narcosis increases our brains are slowly put to sleep." Perhaps this is what's happening during what Stopnitzky beautifully refers to as "psychological softness."

"When breathing compressed air," Sawatzky states, "the functions of the brain are activated, imagination is lively, thoughts have a peculiar charm, and in some persons, symptoms of intoxication are present . . . This narcotic effect [is] due to the raised partial pressure of nitrogen . . . Narcosis is also an effect of 'narcotic drugs' like morphine, demerol, codeine and heroin to name a few . . . The signs and symptoms of narcosis are similar to alcohol. They include laughter, excitement, euphoria, over-confidence, terror, panic, impaired manual dexterity, idea fixation, decreased perception, hallucinations, stupor, and unconsciousness. These signs and symptoms are in rough order of onset and represent the brain slowly being shut down."

Different people react differently to similar addictions, of course, as Sawatzky must acknowledge: "The first part of brain function to be suppressed is the conscious control on behaviour . . . I had the opportunity to be a research subject on dives to 300 fsw [feet of seawater] . . . I will never forget my experience on the first dive. As we were passing 200 fsw I remember thinking the experience equated to the 'maximum effects I'd ever experienced from alcohol.' At 250 fsw I was thinking that 'if this is what street drugs are like, I understand why people get addicted.' At 300 fsw I was in 'nirvana.' Because of the increased gas density I could hear the blood being pumped through my ears. I could quite happily have sat and listened to the sound until I died."

"We're gonna hit the bottom," Stopnitzky croons again, dancing on the marina dock, gently tapping *Fangtooth*'s hull with a squat metal hammer. I will not muster the courage to board the sub, but I have watched videos that show its interior. The joystick that steers *Fangtooth* is labeled with one of those retro label-maker strips, STEERING THINGY. A little black box next to it is labeled ELECTRO THINGY. Next to a gauge measuring $CO_2$ are two canisters housed in what looks like an antique wine rack, also retro-labeled SUPPLIMENTAL OXYGEN, perhaps purposely misspelled. EMERGENCY INTAKE CONTROL, reads the label over a little yellow wheel. COMPLETE FLOOD, reads the label over a porthole. "Of course, you could die," Stopnitzky acknowledges. "I mean, you're in a steel capsule in the depths. Like, you could conceivably die."

She says she's begun to develop "really extreme nerve pain," during which the entire right side of her body seizes, but dismisses it as soon as she brings it up. "I'm really concerned," she says, "but I don't want to miss out on diving. So I force myself to do it, and [when I do] I come out 100 percent cured.

"[Because] when you go on a sub dive," she says, "there are all of these gelatinous creatures that have absolutely nothing in common with anything you've ever seen before . . . And it's utterly captivating to see these things that are truly—they're aliens on our own planet." They are purple jellyfish with feather-legs undulating within a whorl of yellow sea crumbs. They are miniature Milky Ways threaded by thumb-size animatronic umbrellas with white transparent shells, their organs glowing and licking like little campfires inside. They are blue phosphorescent string beans with a thousand leg-fins, and they are orange orchids that move by flapping their petals. They are worms strung with mite-size tea lights. They are ever-flowing wedding gowns, their innards shaped like tiny humans with lightbulb heads. They are the most beautiful ghosts in our world.

Regarding the trajectory of her life, Stopnitzky says, "I'm just kind of floating." She boards her sub, closes the hatch, and a double-crested cormorant alights on *Fangtooth*'s conning tower, centers itself like a candle on a birthday cake. In many world mythologies, the cormorant is a symbol of nobility and indulgence. It is a talisman for fishermen, the harbinger of a bountiful catch. Fishermen once trapped the bird and put a snare around its throat, affixing the poor thing to the masts of their ships, so the fish would come. In Greek mythology, a sea nymph, disguised as a cormorant, saves Ulysses from drowning after a storm wrecks his boat. In Old Norse mythology, the cormorant is said to carry messages—especially warnings—from the land of the dead to the living. In Danish lore, the dead can visit the homes they once inhabited when alive only in the form of a cormorant. In rural Old England, a cormorant on a headstone meant the dead would rise, possess the children of the village, and compel them to murder the adults, and so in order to prevent this from happening, midnight mobs were formed and charged with smothering the children in their sleep.

Across cultures, the feathers of the cormorant are considered to be so beautiful and irresistible that young men would adorn themselves with them when initiating courtships with young women—a proto–Axe Body Spray. A cormorant perched on a church steeple is bad luck. A cormorant at the windowsill of a dying person means the person's pain will go away.

Atop *Fangtooth*, the cormorant lifts one foot, then the other, as if the steel is hot. It remains silent, and doesn't make the deep guttural grunt for which it's often ridiculed among ornithologists for sounding like an oinking pig. In its mouth, something small and white is trapped and twitching. It can't be a butterfly, but it sure looks like it. I wonder if Stopnitzky knows the bird is there; if she can sense it. The cormorant stretches its long black wings and opens its beak. Whatever was in its mouth takes to the air, makes its escape. It circles over *Fangtooth* like a halo. The bird picks a feather from its fire-orange throat patch as the sub's engine rumbles to life.

# 18

OLE STOBBE PUT DOWN his cell phone on the nightstand and sat on Kim's side of the bed, his toes tapping the floor. It was the middle of the night, but it felt like the middle of something bigger. Something more encompassing and lasting. The phone's screen went dark, taking with it the photo Kim took of the distant windmills, the image diving deep into the shadowy code of the phone. As he waited for the police and prepared his heart for their forthcoming knock on the door, Ole touched his cheeks. The sound of his stubble crushing itself against his palms seemed deafening. He buried his face in his hands. When he dug it up, lifted his face again to the cruel silence of the apartment, it didn't feel like the same face.

≈

Madsen was surprised his heart wasn't racing. He put down the screwdriver on the counter in between the plates of chocolate chip and oatmeal raisin cookies. *Nautilus* moaned through the depths of Køge Bay. It was well past midnight, and he wondered if he would get any sleep. *Nautilus* lazed through the strait, northwest of the Øresund Bridge, the impressive eight-kilometer-long tether between Denmark and Sweden. In the years to come, the Danes and Swedes would alternately close their sides to stem the "migrant influx," and to help solidify a coronavirus lockdown. People would be tackled and arrested for trying to cross. Locals would liken the new border restrictions to "going several steps back in time." An official working for the Danish commuter association would compare the bridge's "symbolic identity" to that of the Berlin Wall.

But now, in the wee hours of Friday, August 11, 2017, only a few cars freely crossed, though their headlights did not reach Madsen's submarine. A merchant ship, though, did spot *Nautilus,* but thought little of it, the crew on board having heard of Madsen's antics and experiments and minor local celebrity.

Though Ole, back at his and Kim's apartment, gave his exhausted and anxious statements to the police, and then, in turn, to the Navy (to whom the police suggested he make an additional report), *Nautilus* had on board no satellite tracking devices, and the authorities—as any depth-obsessed DIY submarine fabricator might ensure—had no way of locating or contacting the vessel. Madsen rinsed the screwdriver under the faucet, surveyed the hull, and regarded the futility of this action. He looked upward to the tower, the 70-kilogram hatch, and began to stitch together his story.

He considered what he would have to do. With his tools, he fiddled with the ballast tank, began to sabotage its healthy functioning, which allows the submarine to submerge without flooding, as the tank holds an amount of water sufficient to increase the mass and density of the vessel. But something stopped him from completing the sabotage, at least for now. A wormy anxiety crept into his chest, then ran down his arms into his hands, which popped open, dropping his tools. They gonged to the red metal floor, and Madsen stopped their rolling with the toe of his boot. He stooped to pick them up and felt tired, suddenly unable to keep his eyes open. Madsen: the architect of his dreams, which he dragged forth, kicking and screaming, into waking life.

Waves of nausea washed over him, then dissipated. He felt cold inside, a steely dread creeping into his heart. The dread settled in, and Madsen knew it wouldn't take long for him to grow accustomed to it, comfortable with it. He even considered—right there on *Nautilus*—greeting it. Madsen thought of his wife, S. Though he predominantly used his workshop as his living space, and S. occupied a separate apartment in Copenhagen, the couple often texted each other about the goofy behavior of their three cats. From the get-go, at Madsen's urging, S. agreed to an "open relationship," allowing Madsen to indulge his S&M fantasies with other women at sex clubs and at private sex parties.

Their marriage was stressed not only by the separate living spaces and Madsen's insistence on his particular breed of "openness," but also by a lack of cash flow, as Madsen commandeered much of S.'s income and devoted it to his rocket and submarine projects. S., who would leave Madsen a few days after his fateful voyage with Kim Wall, and who would request of the press a cloak of anonymity (including the obscuring of her name), would also claim that their bank account was so meager that Madsen didn't own any clothes besides one pair of slacks and a button-down shirt (his "sex party" outfit), and a few workshop "boiler suits," which he would wear for weeks at a time.

He shifted in one such boiler suit there on *Nautilus*, swallowing hard, breathing out his nose, greeting his dread. He bit his lower lip but stopped before the skin broke. From the roomy kangaroo pouch of the boiler suit, he pulled his cell phone and texted S. His hands were shaking, and he watched his thumbs shudder over the letters as he typed, "I am on an adventure on the *Nautilus*. All is well. Sailing in calm seas and moonlight. Not diving. Kisses and hugs to the cats." His chewed-down thumbnails, stained a rusty red, clicked on the screen as he typed.

*All is well,* Madsen repeated in his head, tapping his boot on the steel floor of his sub, where the sound echoed like church bells, or alarm bells, called *Amen, amen, mayday, mayday, mayday.* He realized this text message was a farewell to his wife, sure, but also to everything he had ever known, or been. "I realized there was nothing left of the world I was living in," he would later say. "I'm done as 'Rocket' Madsen." He replaced the cell phone into his pouch, gazed at his own navel, and eulogized himself.

He assessed the sharpness of his tools. He contemplated suicide, held the tip of a sharpened nineteen-inch-long screwdriver to his neck, and growled until he laughed or, at least, forced a sound he could mistake for laughter. He had a sudden but fleeting craving for peanut butter. "I was in a suicidal psychosis, and I had no more plans in this world other than to sink the *Nautilus*," he would later claim. "Yes. A fitting end for Peter Madsen would be on board the *Nautilus*. But then, I gained strength."

He took the screwdriver from his throat. He put it on the counter. He stared at the cookies, tried to decode some sort of answer from the arrangement of the chocolate chips. He wanted so badly to die, to live, to build another sub—a bigger and better sub! He wanted people to love him for this, and he wanted to lord over the people who loved him for this. But more than that, he wanted to sleep. "I was so tired and exhausted," he would later report. He plucked a gray blanket from one of the bunks. A pillow. These, he placed on the metal floor, and lay down upon them, next to the stilled body of Kim Wall.

# 19

Madsen's friend, Christoffer Meyer, would later express his surprise. "I have no figures on how many hours I have spent with [Peter Madsen], how many discussions, both philosophical, technical, and personal, we have had, and at no time have I seen this . . . This is not to say that he has not had it in him. Because he has . . . And the big question for me is whether he was even aware that he contained it. I do not think so. But that does not in any way excuse what has happened . . . [Still], one cannot just extinguish one's friendly feelings towards another human being and say, 'That was it' . . . Will I be able to [look him in the eye] in the future? I cannot answer that. . . . I've lost a good friend."

He would consider the images of Madsen and Kim Wall on board *Nautilus* as it sailed away into the Port of Copenhagen, how they looked so "happy, waving, completely calm." Meyer would later wish that Madsen's fabricated story about the "terrible" accident—about Kim's insistence on climbing the ladder up to the hatch door to take a few photos, about how Madsen opened the hatch door for her to do so, about how that 70-kilogram door swung shut as *Nautilus* navigated a wave, hit Kim in the head, and knocked her down the ladder—had been true.

If only Madsen's version had been true, Meyer later bemoaned, "it would have been different," and Meyer would not have to suffer "the grief over the fact that [my] very, very good friend was hiding something that [I] did not know." Meyer would finger the sharp corners of his downturned shirt collar and catch himself, and, referencing Ole and Kim's family, admit, "But what have they lost? I have such great compassion for those people. There is no one to go through . . . ," he would say, and cut himself off, go quiet, and, as if he were waiting for that last thought to complete itself, to make some sense or offer some solace, he would turn his starched collar up and down, up and down, making of it some desperate bird trying to take to the sky, even as its feet were cemented to the earth; flapping and flapping, so terribly surprised that, in spite of all that effort, it covered no distance and was no closer to the clouds.

≈

After dozing next to Kim's body for over two hours, Madsen rose, his back cracking. Suicidal feelings jockeyed for position in his chest with feelings of ecstasy, dread, godliness, and an uncontainable joy that compelled him to cry out in excited whimpers, as he finished undermining the functionality of the ballast tank. He closed his eyes and took deep breaths, in and out, in and out. He opened his eyes and watched his belly inflate and deflate. He ran his hands over himself as he walked to the crackling radio.

Ole, at his and Kim's apartment, had been repeating himself to a series of police and naval officers through the night, and when the sky brightened and turned milky, he offered to make the officers some coffee. His exhaustion did little to mitigate his worry. Absentmindedly, he added too much water to the hopper and the coffee was weak, and he and the few remaining officers sat around the kitchen table drinking the weak coffee. Everyone was so tired, and the officers were wearing yesterday's clothes, and Ole took his coffee in small sips. Beijing began to feel like a dream, the details of which slowly evaporated with each passing minute, some apparitional life raft floating farther and farther from shore, eventually to be usurped by the oblivion of the horizon. A ray of sun speared the kitchen window, reflected from the dark surface of the coffee, and Ole's mug burst with light. One of the officers' phones vibrated. The officer stated his name, paused, nodded, looked at Ole and said, "Okay." Ole felt his stomach drop. But then the officer shook his head and smiled and winked, and all at once, it seemed, Beijing became real again, and the world filled out and brightened and grounded itself, and his petty worries pertaining to their big move blissfully reasserted themselves.

It was 10:55 a.m., Friday, August 11, 2017, 61 degrees Fahrenheit, 86 percent humidity, barometric pressure of 30 Hg. The wind was coming out of the north at 8 mph, which means that when the officers and Ole stepped from the apartment, their hair was barely aroused, and if it lifted from their heads, it did so imperceptibly. The world felt stilled, balmy, liminal, a silent bomb about to go off. Some weather reports would call the clouds that morning *passing*. Others would call the clouds *broken*.

*Nautilus* had been spotted near Drogden Lighthouse in Køge Bay, about eighteen kilometers from Refshaleøen, and appeared to be in trouble. At 10:14 a.m., the lighthouse keeper succeeded in making radio contact with the submarine. "We are having some technical problems," Peter Madsen said calmly over the radio, scratching at his naked torso with his free hand, "and are heading towards port." He zipped himself into an olive-green jumpsuit, rinsed the blood from his cuticles at the galley sink, and practiced his smiles. As he practiced, his smiles grew wider and wider, until they ceased to resemble smiles at all. His mouth became blank and expressionless, even as it hung open, as if deciding whether to yawn or to roar.

Outside, before getting into the naval officer's truck, the officer relayed the information to Ole: Soon after the lighthouse keeper had spotted the submarine and had spoken with Madsen, Madsen himself appeared, distant in silhouette, poking from the vessel's tower. Though it was impossible to see exactly what he was doing from this distance, he appeared to hunch over, then straighten, hunch over, straighten. Soon he disappeared into the submarine, only to emerge about a minute later from the tower. Again, he hunched and straightened, hunched and straightened. *Nautilus* began to sink, and, in less than thirty seconds, the vessel dropped beneath the surface. Two helicopters and three ships participated in the rescue. Ole chewed his fingernails, waiting for the officer to confirm that Kim was okay, and perhaps the officer recognized this, and smiled reassuringly again, and fast-forwarded the story by loading the message from the Armed Forces Rescue Service on his cell phone, holding the screen to Ole, who felt himself lighten to the point of levitating when he read, "The submarine has been found. Everyone on board is okay."

But by *everyone*, unbeknownst to Ole and the naval officer, the Rescue Service meant only Peter Madsen. Kim, at the time the message was sent, was no longer on board. The Rescue Service sent a crew of four men out to the submarine's wreck site, the gray water arcing and spraying as they sped. Madsen bobbed in the bay, and though his hands and feet were growing numb, the water felt good against his chin, his lips. His heart shuddered against the cold as if restarting, and he indeed felt partially cleansed, reborn, clearheaded. From his vantage, he could see only the bow of the motorboat coming at him, towering over him as if some biblical mirage, the crew of four reaching out to him like phantoms through the spray. He flailed his arms and reached back, and the hands of the men felt good and strong as they grabbed at his sleeves, his elbows, his hands, and, as they lifted him from the brine, reached around to brace his hips and back and thighs, pulling him on board. So much water dripped off him, returning to the sea.

Overhead, the planes whined as they descended and lowered their wheels, the airspace over Køge Bay being part of the primary landing approach to the Copenhagen Airport. In them, passengers tapped their toes on their carry-on luggage, stared out their windows at the gray water, chewed gum, swallowed, braced themselves for the landing. Onshore, the dense cluster of suburbs stretched until it petered out at Stevns Klint, a forty-meter-high white chalk cliff, which contains a sixty-six-million-year-old layer of dark *fiskeler,* or "fish clay." This particular example of *fiskeler* is famed for containing unusually high levels of iridium, the second densest element on earth, and the most corrosion-resistant metal known to humankind, so named for the Greek goddess Iris, the personification of the rainbow. To an international consortium of geologists, physicists, paleontologists, and astronomers, the presence of iridium serves to justify the hypothesis that the worldwide Cretaceous-Paleogene mass extinction event (during which over three-quarters of Earth's plant and animal species were destroyed, and no tetrapod—save for an ectothermic species of sea turtle—weighing over fifty pounds survived, making way for a new dawn of the little things) was initiated by the impact of an asteroid with a diameter of nine miles, more than that of the Martian moon Phobos, and nearly twice the height of Mount Everest. Because the bryozoa chalk in the cliff naturally absorbs shock and is impervious to nuclear weapons, the top secret Cold War–era fortress of Stevnsfortet was built here in 1953, deep into the passageways within the cliff. Still today, defunct and decomposing within the cliff's 1.7-kilometer labyrinth of tunnels, are bygone ammunition depots and cannons, living quarters and command centers, a hospital, a cafeteria, a chapel, a crypt.

Mass extinction, nuclear shock, rainbows, sarcophagi, fortification born of fear: the little motorboat passed it all, pounding the waves as it headed toward Dragør Harbor and the teams of police and press waiting on the dock. On board, slick with the sea mist, the rescue crew asked Madsen how he was feeling, to which he responded with two thumbs up, and, perhaps in reference to *Nautilus,* or the nature of his heart, or perhaps in reference to something much larger and looming—something indistinct but encompassing that, as yet, Madsen could not fully make gel—by yelling over the boat's motor and the wind, "I just have a few things to fix."

Those waiting onshore wiped the cold from their noses, blinked the wet from their eyes. The moored boats creaked around them, their tall masts tracing ellipses onto the sky. Behind them, in the tangle of alleys, along cobblestone streets and within centuries-old yellow-painted houses with bright red roofs, locals ate their late breakfasts, sipped their coffee, their windows still closed against the chill. Dragør (which roughly translates as "the gravel beach where boats are dragged ashore"), adding to its reputation as a busy fishing port, also became known for its agricultural output in the early sixteenth century after King Christian II (aka Christian the Tyrant, architect of the infamous Stockholm Bloodbath, during which the monarch slaughtered eighty-two members of the Swedish nobility after inviting them to a private banquet at his palace, solidifying his capture of and rule over Sweden) paid twenty-four agriculturally innovative farming families from the Netherlands to settle in Denmark and cultivate food for the royal court.

As King Christian continued his campaign of terror, beheading and hanging Swedish bishops and noblemen, burgomasters and magistrates, town councilors and "commoners" (and, in one particularly brutal documented instance, ordering the digging up of the disemboweled corpse of former Swedish regent Sten Sture the Younger and burning it at the stake alongside the body of Sten's young son), the Dutch settlers introduced to Denmark the carrot, which was so beloved by the royals that the jealousy and ire of the local fishermen soon became aroused. The fishermen were known to trek inland from their coastal bungalows in the middle of the night, setting fire to the Dutch encampments and beating the families as they rushed outside to escape the smoke. The Dutch fought back, and the area saw violent clashes that the royal court ignored, so long as they received their set claim of the carrots and the fish. Such violence kept the locals living in peasant conditions until the late nineteenth century, when Dragør's major seafaring prosperity began and never significantly waned.

The adjective most often used to describe the neighborhood in local guides and brochures is *charming*. In spite of tensions that persist between the Danish fishing families and the descendants of the Dutch farmers, Dragør is known today as the place where "the happiest Danes live." Still, few residents smiled as they sipped their coffee, chewed their toast. When the rescue boat reached the harbor, it did so at such a speed that it slammed against the dock, and Peter Madsen nearly fell to his knees. As Madsen stepped from the motorboat to the boat bridge in his clumsy boots, he stumbled over his own feet three times, before a member of the police force offered his hand, and Madsen took it, steadying himself.

# 20

"Peter?" a female journalist called to him as he stumbled onto the dock at Dragør Harbor. "Peter? Peter?" At the third mention of his name, Madsen looked up, bewildered, toward the direction of the voice, trying to discern from whom it was coming. "Are you okay?" the journalist asked. Madsen looked perplexed, and she repeated the question: "Are you okay?" He raised his right thumb and averted his gaze, down to his boot tops, which, though he had earlier scrubbed them clean, still bore round, dime-size stains and splotches. He stared at these and told himself stories. The droplets could easily be mistaken for something else; he knew that. Also staining his boot tops were varieties of oil and fuel.

A member of the rescue team wrapped a golden foil blanket over Madsen's shoulders, and Madsen clutched at it with his left hand, cinching it over his neck like a cape as the press crews rolled their cameras, snapped their photos. Madsen's hair hung in sodden strings over his forehead. The wind picked up speed. As he faced his audience, his heart was screaming on the inside, and he could hardly contain his excitement, but on the outside he was all smiles, all thumbs, all thumbs-up.

Someone passed Madsen a bottle of water and patted him on the back. The foil blanket made a crunching sound. The rescue team coiled lengths of black rope. The wind roared and troubled the microphones attached to the video cameras. Various journalists shouted their questions over the wind, their voices intermingling. Madsen didn't know who to answer first. He appeared confused, and still smiling, and when he drank from his water bottle, he did so rapidly and erratically, as if downing shots of tequila, as if his throat were desperate for the moisture. The water was subterfuge, disguising his nervous swallowing.

One journalist, aware of Ole's missing persons report, inquired about Kim Wall's whereabouts. Madsen laughed and looked down at his boots. He took the foil blanket from around his neck and crumpled it into his left hand, held it balled against his waist. Still, the wind compelled it to crackle as if a live wire. With his right hand, Madsen swirled the water in the bottle. The camera rolled, and Madsen lifted his eyes from his boots, looked right into the lens.

"I am alone," he said, his voice thin but jovial. Perfectly calm. "She was not with me on the submarine when it went down."

"What happened?" another journalist asked.

Madsen continued to speak about Kim. His voice sounded flat and oddly noncommittal, as if he were reciting a half-baked first draft of his story. It sounded ill-rehearsed, and he covered this with additional nods and smiles. "I dropped her off at about 22:30 last night near the Halvandet restaurant," he said, referencing the establishment on the northern tip of Refshaleøen where Ole accompanied their companions to the after-party.

"Last night? Thursday night?" the journalist pressed.

"I have not seen her since," Madsen said. Later, Halvandet's owner, Bo Petersen, a dashing thirtysomething entrepreneur, told police that the area outside his restaurant was surveilled by a CCTV camera. Petersen is the sort of man who—infuriatingly—looks handsome in most any context. When he handed the footage over to the police, for instance, his salt-and-pepper goatee appeared freshly chiseled, with such sharp angles and edges, and such perfect arcs, as if he had shaved with a compass and protractor. His hair—also salt-and-pepper—was slicked back and short on the sides, some corporate version of a pompadour. He smirked as he handed over the video, said it included no images of Madsen and no images of Kim. The eager but tired policeman reached for the tape and looked comparatively rumpled. The sleeve of his uniform was too short and unironed. Petersen's hairdo did not shudder—not once—in the wind.

At the dock's edge, the journalist asked Madsen for confirmation:

"Halvandet?" and Madsen nodded forcefully and repeated himself: "I have not seen her since." He wiped his nose and smiled and smiled into the wind and the cameras.

"Peter?" another journalist called, and Madsen was on a roll now. He tightened his grip on the foil blanket, and the blanket complained.

"I have her contact information stored on my mobile phone," Madsen continued, "which is now in the submarine at the bottom of Køge Bay." And a few reporters giggled, and so did Madsen before licking his lips and sobering his demeanor.

A policeman in a black padded vest and faded blue short-sleeved button-down shirt motioned for Madsen to follow him. The officer's head was shaved to disguise his baldness, and his brown stubble was carefully crafted. His black pants were too snug, and he intermittently tugged the material from his crotch. He wore a chunky black wristwatch that was so tightly fastened it remained motionless as he gestured to Madsen. Before the two men began their purposeful walk away from the dock and the cameras, the officer asked Madsen what other information he had about Kim. Within earshot of his audience of reporters, summoning the bravado of the eccentric genius persona for which he knew he was locally famous, he answered offhandedly, "She's from *Wired* magazine," and lifted the water bottle to his lips and sipped. "I only know her first name," he continued condescendingly, vocally winking at the press. "I don't check the background when there is a journalist calling and saying, 'Can I get an interview?'"

Some of the journalists giggled again, and one clapped their hands as if to applaud the resilience of Peter Madsen's delightfully ornery personality even in the aftermath of *Nautilus*'s sinking. He was their temperamental antihero, their local patron saint of eccentricity. Madsen smirked and sipped and followed the officer, struggling to keep up, as one persistent journalist and his camera crew ran alongside filming. The foil blanket, as Madsen walked, grew louder in the wind, and all parties involved had to bellow so their voices could be discerned.

Madsen and the officer walked over wet gravel and tufts of weeds. Madsen was still wet, his socks saturated, and water squelched in his boots. His hands were chapped, and the troughs between his fingers were pink and itchy. He kept scratching at one hand with the other. The men came to a police car—a white Volkswagen topped with two rotating blue lights, fluorescent yellow reflective paint around the lower part of the frame, and the word POLITI stenciled black in all caps on the hood—parked in the middle of a skinny paved access road amid the gray boat storage warehouses, the sorts fronted with three-story garage doors. The police car's license place—AV 91 083—appeared brighter than the rest of the car, as if fresh from the factory and newly attached. An elderly couple—the man in an overlarge gray sweatshirt and black beret, the woman in a faded sea-green windbreaker, her dyed blonde hair tightly sprayed—stopped their bicycles behind the police car, curious and eavesdropping. The man had attached a blue crate over the rear wheel of his bicycle as a basket. It was empty. The policeman, taller and wider than the two of them, gave them a withering stare as he urged Madsen into the backseat. The couple, unbothered, straddled their bicycles and remained. The policeman stood straight and stiff, and his forearms jumped. He stared at the couple, realized he could not scare them off, and turned to speak to a plainclothes colleague who had arrived on the scene, emerging from a 4x27 yellow cab, which had pulled alongside the police car.

A photographer from the local TV 2 station hurried up the road and set her camera in the soggy grass adjacent to the police car. She zoomed in and focused on Madsen's face in shadow in the backseat. The unedited footage shows Madsen staring straight ahead, not once turning to look out the window either toward the camera or the policemen, his gaze fixed on the front seat headrests, or windshield, or the gray sky and blowing tallgrass beyond the windshield, or the dashboard, or the dashboard clock (which was reading twelve noon), or the dormant speedometer, or the locked steering wheel, or his own reflection—his eyes, his brows, his forehead, his hairline—in the rearview mirror, practicing his blank look. He swallowed, and in the footage, as voices commingle in the distance, hushed and gossipy, and as the wind crackles against the camera's microphone, his wet hair curls forward from his brow, his nose appears square and small, his lips tightly closed. The image shows only his right profile, as he stares and the surrounding voices continue to mutter.

TV 2's journalist caught up to her cameraperson, and she approached the police car, knocked on the backseat window. Madsen jumped as if having been roused from a dream, and he pulled the foil blanket more tightly around him. Through the closed window, inexplicably—as if he had lost sense of this reality—he asked the journalist, "What kind of thing are we in?" referencing perhaps the police car, or the entire present moment.

The journalist, baffled, answered, "The police just need to keep track of what's been going on."

"Uh-huh," Madsen said.

"Are you feeling okay?" asked the journalist.

"I'm fine," Madsen said. "Watching the *Nautilus* go down was extreme, but that was fine enough. The night sailing: it was a practice trip."

"Practice for what?" the journalist asked, but Madsen didn't reply. He squeezed his water bottle over and over, and the plastic made that crackling sound. With his other hand, he clutched the foil blanket, balled it over his throat. It seemed as if he needed his hands to be full, and working, gripping the bottle and the blanket.

His hands rounded and clawed, he cleared his throat and said, "I played with some things."

The journalist cocked her head to one side, as if prompting Madsen to elaborate, and he did: "Then an error occurred on her ballast tank. It wasn't very serious until I tried to repair it. Then it became very serious. Then it took about thirty seconds for *Nautilus* to sink."

Madsen dropped his eyes and lifted them, dropped his eyes and lifted them. His hands were clinging to anything they could—the plastic bottle, the foul blanket. His hands were shaking as if in anger or horror. Maybe he could no longer tell one from the other, and he felt dizzy, and his vision blurred. He began to conflate Kim Wall with *Nautilus*. He called them both *she*. They morphed together, became a hybrid but singular body, a manifestation—if only in Madsen's mind—of the Danish concept of *grænseoverskridende*, or "the inappropriate or perverse transgression of boundaries," which can be applied, of course, to Madsen's own desperate need to shun gravity—to rise and to sink, to be rocket and sub, bird and fish, inventor and murderer. To be, via his constructions and adventures, a storyteller who needed to free himself from the confines of those very stories he perpetuated; who empowered himself, however brutally, to achieve that freedom. To float, to bob, to hover, and, so hovering, to bury at sea.

Another taxi pulled up, tires spitting gravel. A male journalist in aviator sunglasses leapt from the backseat and rushed to the police car, stepping in front of the female journalist, crying, "Peter! Peter! A few words?" through the backseat window.

Madsen shook his head.

"Please, Peter," the male journalist begged, his sunglasses slipping down his nose. "Where's your partner?"

Madsen again envisioned Kim Wall fused with the carapace of *Nautilus*. He pictured her face as a porthole, smiling and then screaming, and then not doing anything at all. He swallowed, and the porthole went foggy with temperature and depth. The bottle and blanket made their strange percussion in his hands. "Well," he answered, "there was no one but me on that trip there. She wasn't there."

The police officer rushed over to the car and tried to shoo the journalists away. "You can talk to him afterwards," he said in a scolding voice, and Madsen surely wondered: After what exactly? He could not know, there in the backseat, squeezing and squeezing, that it would be another several days until he would speak to the press again. The male journalist complied, but the female journalist lingered, and the policeman kicked the dirt from one black shoe with his other black shoe. At some point during these encounters, everyone involved wiped their noses on their sleeves.

Off camera, a wave must have slammed the Dragør Harbor jetty because the already cloudy sky bloomed with mist, and an awful, echoing digestive sound overtook the muttering voices. So many varieties of seaweed danced, tentacular shadows in the murky brine—the bullwhip kelp, the creephorn, the dead man's bootlaces. When the sea settled, so did the seaweed, onto the coral beds and emptied shells, a pile of spines that once lurked inside fish, the carapaces of long-dead crustaceans, and the body of a newly dead woman.

As the sea crashed against the jetty of Dragør Harbor, the journalist asked Madsen through the closed window of the police car, "How are you, Peter?"

Peter Madsen did not look at the journalist, but continued staring straight ahead when he answered, from the backseat, "I'm fine." Then he laughed a little. He laughed a little and said, "I'm a little sad."

# 21

The "psubs-mailist" is rife with blind alleys and digressions, in-jokes and infighting, scrubbed attachments, redactions, coded language, and intricate jargon bordering on the fetishistic. It is a byzantine network of online chats, both argumentative and supportive, and combing through it all is akin to ice-skating along a Möbius strip while trying not to drop an armful of exotic tools and submarine parts, the function of which remain elusive in the murk. This community has developed its own rivalries and allegiances. In short, it's awfully entertaining—if often bewildering—to navigate.

The cliques and factions and subfactions can be reminiscent of a high school social network. The newcomers to the group are often treated by the others as seniors treat freshmen, with, as one poster called it, "total bluntness," shutting down conversations "by a type of intimidation, on subject matter some were tired of discussing and/or not interesting in."

"Neophyte PSUBers often take as gospel any information found here on the PSUB list and can lead to misconceptions on their part in the future . . . some of which may be dangerous if not downright deadly . . . A number of us have to waste a good deal of time better spent on other endeavors dispelling misinformation," warned one member, to which another responded, championing further exclusivity, "Perhaps there is a need for a new forum for a select group of individuals that don't wish to deal with teaching and/or reading what amateurs are talking about, and want only very serious factually accurate post in there [*sic*] group. To join this group one would need to have done more then 500 dives in a sub worth more then $100,000."

In friendlier, quieter moments, though, it's just a bunch of guys getting nerdy about submarine parts—consulting with one another over propellers and ballast tanks and acrylic windows. The only woman who seemed to contribute to the mail list was Shanee Stopnitzky, and she did so only briefly in March 2018, announcing herself to the group with the heading "newbie looking for core starter info," and soliciting advice on the retrofitting of *Fangtooth*. Stopnitzky's presence injected new life into the chat space, the men excitedly engaging either their mansplainy or lecherous or fatherly tendencies. Stopnitzky used punctuation marks to "draw" out a seascape beneath her signature—tildes and carets evoking a wavy ocean surface, fish fashioned from greater-than and less-than signs. "It would be also great if there is anybody out there who is particularly enthusiastic about answering many newbie questions because I will have a lot and don't want to annoy the whole group all the time," she wrote in her first post. She signed her messages, "Best fishes," and closed out with Primo Levi quotations. Hank Pronk and PSUBS founder Jon Wallace were among those to enthusiastically welcome Stopnitzky to their community and to provide advice. Others were eager to give Stopnitzky their private contact information. ("If you live close enough i can do it all for you," one wrote.)

Many times, though, these guys simply negotiate deals over sub parts, buying from and selling to one another. A disproportionate number of them quote Yoda in their signatures. But whenever Peter Madsen raised his head into a chat thread, it was as if a rock had been thrown into the pond.

Before murdering Kim Wall, Madsen was a fixture on the seemingly bottomless pit of the PSUBS-MAILIST, on which he wrote, a decade before the crime, in August 2007, "One major factor of motivation for me—in submarine construction—is its ability to make me personally virtually government uncontrollable." His post generated a plethora of supportive and like-minded responses.

Madsen, Hank Pronk, and others often communicated on the same PSUBS threads—about submarine parts and adventures, and accidents—and such correspondences often bore energetic swagger and fanaticism. They developed, in part, their own language, and, in turn, their own worlds, and hierarchies, and sets of values. These men, ensconced in their cloisters of submersible and community, cared deeply for one another via their shared obsession, even if such care tended to border on the cultish, and to the exclusion of the remainder of the world.

When, in the late 1980s, drug smugglers began to transport their wares via private submarines in order to avoid radar detection (wrapping the steel hull in Kevlar in order to make it more difficult to trace), the personal submersible community came under fire. The practice gained traction in the mid-2000s. Most everyone who owned a home-built submarine generated a measure of suspicion—either as a participant in the smuggling rings, or as someone who would potentially sell an amateur sub to these syndicates. Some in the PSUBS group, while not condoning the actions of the smugglers, begrudgingly admired their handiwork. "Something like that would have taken a year or so in a modern shop," said one of the members when consulted about a confiscated "narco sub." "Imagine doing it out in the boonies with the mosquitoes and vermin!"

In 2007, members of the PSUBS community were allegedly accosted and interrogated by mysterious officials who claimed to work for the FBI. This did little to quell the systemic distrust of authority and the government that already shaped many of the conversations among the group. There were rumors that guys like Jon Wallace would have to disclose their hobby to the government and report on the nuances of the submersibles' designs, including any design flaws. Wallace responded with a fiery manifesto on the PSUBS online chat forum, earning him the cheers of his fellows.

"I do not believe PSUBS as an organization has an ethical obligation to inform anyone, unsolicited, that their design/construction techniques are flawed," Wallace wrote. "You will have a difficult time convincing me that they are acting in the best interest of the underwater community as a whole . . . [The Marine Technology Society, American Bureau of Shipping, 'Coast Guard and/or government' to which PSUBS would have to answer should such regulation come to pass] considered PSUBS a bunch of nutjobs, weekend warriors, who were going to endanger the entire industry because we were all idiots with no discipline … and would like nothing more than if PSUBS just disappeared."

"Once they get there [sic] hooks in, they will regulate it to death!" another PSUBS member cried.

Wallace tried to both soothe and agitate his members. "No entity is in a position to better represent the unique issues of submarines used for personal purposes than PSUBS," he said, before issuing a delusional demand: "PSUBS will fight to protect our own interests, specifically that the Coast Guard adopt PSUBS rules and regulations for personal submarines."

Only Hank Pronk, repeatedly mentioning that he was speaking as a Canadian, seemed to feel uncomfortable with his community's fiery reaction to a rumor of regulation. "I think we might be overreacting if we think Psubs are dead with rules," he said. "I had a look at Canadian rules for [amateur] built aircraft. It is not a big deal. It would not be the end of Psubs if we had to follow these rules." Pronk aroused the ire of the group, which, though he had spent years cultivating the respect of his peers, responded with an irate pile-on.

"If you go down the 'Rules' path, where does it end! In general society doesn't care too much if you kill yourself," wrote one.

"I cringe at the thought of any government regulation of private submarines!" wrote another.

Pronk had to walk back his comments to be reembraced by the community. "Not that I want rules—that is the last thing I want," he said. "Seems, poking a beehive with a stick is bad!"

Madsen was a controversial and outspoken fixture on the PSUBS chats, admired for his engineering and design prowess and the "greatness" of his pre-*Nautilus* subs, *Kraka* and *Freya,* but also generating a wariness for his occasionally extremist and unhinged rants, in which he was frequently fixated on the Nazis. Apparently he entertained the rumor that at the end of World War II both Adolf Hitler and Martin Bormann faked their deaths with the help of top Gestapo officials. They secretly fled to Argentina on what has become, to such conspiracy theorists, a legendary submarine, carrying falsified Vatican passports and masquerading as reverends named Gomez. The rumor postulates that Hitler and Bormann continued to possess economic influence in their safe haven, quietly wielding their power more via "talking" than "doing," ensuring that the Third Reich would survive underground in Mafia-like fashion, waiting for the perfect future moment in which to rise again like a Whac-A-Mole. "I wish to take no part in this stuf about who is bad and who is good," Madsen once wrote on a PSUBS thread that engaged the difference between "doers" and "talkers" within the personal submersible community. "Most really important people are in fact talkers [too]," he continued, before honoring said importance with an odd equivalency, "J F Kennedy—Churchill—Adolf Hitler—Werner von Braun—mostly talked . . . along with Josef Stalin and Albert Einstein."

In August 2007, Madsen attempted to sell *Kraka* to another enthusiast though the PSUBS site, and when it took longer than he had expected, he expressed surprise and disappointment that no would-be drug traffickers had yet contacted him, which prompted some reactionary responses.

"Studies of blue whales can be made by military hydrophones," one poster wrote on the thread, "so you can track all individuals of the population of a whole ocean at any time. So if you can track a blue whale over an ocean—how difficult would it be to track a trafficker submarine running a diesel!!!"

Oddly enough, this poster, losing himself to the excitement over his own knowledge, perhaps, went on to describe *how* a sub could avoid such detection. "So IF you smuggle by boat," the poster continued, "the key factor is to [camouflage] among the thousands of surface boats a sonar operator has on its screen at any time."

Madsen responded, "If the drug smugglers did it right—with real uboats outfitted with diesel electric propulsion—and using snorkels—frankly the transit from south American waters to US waters would be very safe for them . . . Problems arise when on loading and off loading the cargo—but even this has a technical solution . . . but like the in the case of the Nazi's I am impressed with the audacity they display by trying to build an functional long range smuggling boat."

When Madsen received backlash to his post, he responded with an ultra-long and predominantly boring rant, rife with cryptic platitudes and half-baked connections, attempting to "educate" and vaguely threaten his peers. "There is no way to do something that is wrong in the right way," Madsen stated. "War is about deception, cheating . . . Its not limited to the battlefield—its start when you make up your brand . . . Part of that is to tell anybody that you can smell what they are thinking about . . . Frankly—with the right kind of boat—it would be very easy [to smuggle]." It seems as if Madsen's excitability or anger got the better of him as he typed, as a preponderance of typos ensued. "How wonderfull a world we live in," he continued. "What stunnes me is how much you belive in authorities . . . I have sailed surfaced down the Copenhagen habor with a live, Sidewider size missile on my back deck—everybody smiled and took pictures . . . People—are exceptinally stupid—until they are hit by fact—as in the case of september 11 th . . . I have—closed to the HMS Invincible—a Royal Navy aircraft carrier—with none noticing the uboat until I said 'hello!' up the hatch. They were not expecting a midget submarine . . . ergo I could have . . . 100% guarentied . . . dropped a mine on her side—and retreated safe to enjoy the sight of the Invincibles sinking."

Madsen's rant is actually much, much longer, and the correspondence spiraled into the wee hours of the morning, and into the next day, across numerous times zones, the antic fingers of men with submarines in their garages typing and typing, waiting perhaps for the FBI to knock at their doors, Madsen furious and desperate to prove his dominance.

Madsen's disagreement with his peers over the futility of drug trafficking via home-built submarine and his Nazi sympathizing soon took on an even more ominous tone, as he personally targeted his PSUBS detractors. At 2:46 in the morning, Denmark time, he continued, "Take a look outside in your swimming pool . . . look close . . . see that's my periscope on UC-3 taking high definition photos of you . . . as you write your answer . . . Might I add—that I have been closing in Freya to a German missile armed destroyer in Copenhagen habor—being detected only at distance of three feet—by German sailors smiling and taking pictures . . . Regards, Peter."

One poster got in Madsen's good graces not by disagreeing with his assessment of the Nazis, but by simply trying to separate the actions of the Nazis (whom he said he did not admire) from the actions of the German Navy (which he did admire) operating under the Nazis. This poster went on to gush over Madsen's sub *Kraka,* calling it "very beautiful" and wondering about the cost of building it.

Madsen, of course, felt the need to prove that he was the smartest guy on the thread. "In writing 'Nazi' submarines I simply tried to be polite. In Germany you will typically say the Nazi's did this and that—When the German guide in Concentration Camp Sachenhausen [*sic*] shows the tourists piles of human hair and shoes—she does not say—this is 'German' industrialized murder—but that the 'Nazi's' did this . . . so that she avoids being identified with it. Today—I saw a picture on wiki—showing a person in a Luftwaffe uniform lying in a bathtub with icebloks around him. He's is German—but unfortunately also a Jew. With him is two Luftwaffe medical officers—writing down his reactions on paper as he is slowly freezing to death. They measure time vs. body temperature and takes notes when he loses continence and when the heart finaly stops . . . The submarines of both US, Soviet and German navies killed sailors, civilians and soldiers by the thousand in much the same way . . . Its not a German, Soviet or US specialty. Submarines did their gruesome part—and may do it again . . . fantastic machines."

To which another rightfully exasperated and perhaps horrified member of the group responded, "What, exactly, does any of this have to do with psubs?" And maybe Madsen had finally worn himself out and went to sleep, because, amazingly, he did not respond further.

But many others did, still eager to chat about the Nazis and express deep distrust in the government, the media, and other institutions that tend to shape and control public opinion, practice, and life on the surface. It is this zealous distrust that has, in part, sent many amateur submersible builders underwater where they can be immune to the bureaucratic rigmarole and information delivery systems of the surface dwellers. Things are simpler at depth if one only has to control the sub in which one is the sole inhabitant and sole demographic. At depth, when one's fate is stitched to the tight vessel in which one is enwombed, then the vessel, in this way, becomes the world. The sub becomes the globe, the nest. Many posters—some correctly, it would seem—became paranoid that they were being surveilled.

"Last year, ▓▓▓▓▓▓▓▓▓▓▓▓▓ and I took his psub, Vindicator, out for a dive in a lake in Arkansas," one member wrote. "As expected, towing the sub from Louisiana to Arkansas generated quite a bit of fascination and interest from drivers and from those at the lake. After a weekend of diving and returning to our respective homes, I got an email from ▓▓▓▓ saying that shortly after arriving home, he answered the front door to two FBI agents (I don't know if they were wearing dark glasses). They had been contacted by someone up at the lake about a suspicious sub and were interested in what we were doing with this private sub . . . From that point forward, I have been careful of what I say in this forum as I now assume all psub emails are parsed for trigger words by some mainframe in Washington DC."

To this issue of avoiding trigger words, another member defiantly responded, "jihad bomb assassinate security smuggle Bush disguise terrorism target DEA FBI NSA circumvent mole hidden camouflage detonate airliner crowd demonstration identify damage genocide Satan CSIS explosives training officer martyr fascists . . . Now that I have your attention, welcome to the list guys! Who knows, maybe one of you will actually take an interest in PSUBS as a hobby."

And the messages continued, citing the ways in which we've never fully moved on from the Hoover era of American history; making references to bomb warehouses and mysterious men showing up at dive sites. Whether such encounters with the FBI were true or not, or exaggerated or not, even those who never endured such encounters tumbled down the rabbit hole of the thread, and it spiraled into the wee hours of many a morning.

"Welcome to the future big brother is watching you," one wrote in the middle of the night. "Let me tell you what some of you guys have to expect because it already happened to me. You will get a house search by a team of 10 drug agency agents . . . What will be checked for is: drugs, arms, bomb stuff. You will be interviewed how you earn your money, how much value is your property . . . This will happen months after your neighbours warned you that a suspicious Mr. Smith—is asking about you what you do—where you live who is with you—when you leave the house . . . etc . . . Keep your mrs calm do not lock your door—they would take it as 'crime in progress' so get violent—colaborate with authorities in the way peter suggests—take it with humour. Finally they are just doing their job. Sniffing submariner, rocket, and airplane community is part of it. My educated guess of next on the list: peter, sorry you earned it—"

The Peter to whom he was referring was, of course, Peter Madsen, and Madsen's original and harrowing message on the subject contained such nuggets as, "I don't like goverments and I don't like the rules and that prevent me from building rigid airships and manned spacecrafts or airplanes or anything else . . . even operate personal submarines . . . I used to work with amateur experimental rocketry before I made the shift to manned submersibles. That meens working with solid, hybrid or liquid propellant rocket motors . . . Building these things from scratch . . . The very best was to be on TV—since no goverment or police official could believe that a terrorist group would develop weapons of mass destruction on live TV . . . Essentially—intelligence organizations need to be told exactly where to look. Lack of imagination is their biggest problem. Lets help them."

Given what we now know, the final line of Madsen's message carries with it a smug and chilling resonance. But at the time, it begat only a flurry of responses with such all-caps statements as "MR. FED-MAN, PLEASE STAY AWAY!!!" and "ARE WE LAWYERS OR HOBBYIST?" To the latter question, Madsen again felt compelled to cryptically weigh in: "Harassing, ramming or torpedoing whalers is possibly quite legal."

≈

When Madsen announced to the PSUBS chat group later that August 2007 that he had finished *UC3 Nautilus,* he did so under a new username. He went from being "Peter Madsen" to (yes, in all caps) UC3 NAUTILUS. He so identified with his new "masterpiece" that, in this way, he became his sub, or at least betrothed to it. He took its name.

"The Supper [sic] Secret UC3 Pictures Are Finally Available for Viewing," the newly christened Madsen titled the thread. Like most megalomaniacs, modesty was never his strong suit, and his superiority complex and refusal to concede any argument to his colleagues had gained him a loyal following on the site. Even those who took issue with some of his more outlandish posts about Nazis often begrudgingly admired his technical and design prowess.

Most of the responses to Madsen's news were some variation on "Woo Hooooo." Some others, following their obligatory congratulations, got a little strange, though. "She's a magnificent boat and well worth the effort and sacrifices you have made to bring her into being. You will have to sail her across the Atlantic for the next P-subs convention. ☺" one began, before inexplicably pivoting: "If you can install a few torpedo tubes I would like to hire her to deal with some Japanese whalers that will be hunting Humpbacks this season! Warm regards."

"I love the build!" wrote Hank Pronk, and many others similarly gushed, asking question after question about *Nautilus* as if a bunch of overexcited kindergartners. Madsen responded coolly, tossing his subjects no crumbs: "First . . . ships are build and turned operational at in a bit different way than TV sets or washing machines . . . UC3 is a whole new design . . . the ultimate expression of knowhow and experience." At one point, he actually referred to himself in the third person as "mr. Madsen, Master and Comander of his homemade submarine flotilla."

"You are a good captain, I would sail with you any day!" one of the Master and Commander's disciples exclaimed.

Ten years later, in 2017, news of *Nautilus*'s sinking lit up the PSUBS chat group, the twitchy urgency of their messages serving as an odd archive of the real-time unfolding of events.

By 2017, Kim Wall had traveled through South India, Israel, Palestine, Cuba, Haiti, North Korea, Uganda, Lebanon. She had studied at the London School of Economics and Political Science, supplementing her income by working at bakeries. She had lost her grandmother. She had plans to plant a chestnut tree in the garden of her first real home. She had published her work in *The New York Times, The Huffington Post, Time, The Atlantic, The Guardian*. She had written about vampires and voodoo, hustlers and hackers, atomic bombs and feminism. She had reread all the *Harry Potter* books, dog-eared on so many nightstands around the world. She howled back at howler monkeys and they howled back at her. She danced in a desert. She slept in a rowboat. She fell in love. She ate ice cream.

The first message on the PSUBS chat about the sinking of *Nautilus* was posted by one of the group's frequent contributors on Friday, August 11, the morning after Madsen murdered Kim, soon after Madsen was brought ashore; soon after he postured for the journalists and was taken across the dock to the backseat of the police car, wrapped in that foil blanket. At first, the PSUBS community's concerns were with the loss of the vessel itself. They mourned *Nautilus*.

"Well, sad news," the first post reads, "as the UC3 Nautilus sank on a trip. Fortunately, nobody was hurt." The member goes on to speculate that the sinking resulted from a fault in *Nautilus*'s ballast system. "It is a fascinating hobby as long as everything works out," the post concludes.

Until further information was released, the subsequent posts show a community rallying together around one of their own, toasting a once-great submersible.

"Wow. So sorry to hear."

"I can't imagine how disheartening that must be. Particularly with a boat of that magnitude. Peter, I'm just glad you and your companion are safe."

The grisly details would soon be revealed, but before they were, Hank Pronk was one of the first to respond to the news of *Nautilus*'s sinking. "I am real sorry to hear that! I presume a recovery plan is in the works? How deep is it sitting?"

Around that moment, in Denmark, Madsen, from the backseat of the police car, told a journalist, "I'm fine." Then he laughed a little. He laughed a little and said, "I'm a little sad."

Immediately following Pronk's worried post, another member of the PSUBS group, having done some further research, changed the tone: "So, I've got to ask about this incident because some news reports are saying that a woman died on the sub. Does anybody know the true story about what happened?" The responses accumulated in quick succession.

"He is being held on suspicion of murder!"

"Getting crazier by the minute. I read the boat is in only 7m, which makes no sense if someone was trying to sink it to hide evidence. I hope they get inside to resolve this fast."

"Not looking good; He took the reporter out, & didn't return . . ."

Scott Waters, one of the more respected and accomplished members of the PSUBS community, weighed in from his home base in Kansas. "My electrical engineer is in Denmark and knows him well and is currently there," he typed that night. "The submarine sunk in very shallow water (like 7 meters). There was a female news reporter on board. After they were rescued he took her home then she was reported missing. Everything is currently under investigation."

Around that moment, about a kilometer away from the police car in which Peter Madsen sat, Ole Stobbe shivered in his windbreaker and paced.

"Thanks for the update!" another PSUBS member responded to Waters. "By tomorrow, everything should hopefully be clearer. I hope everybody here is smart enough not to believe what the news is saying until the official police report gets posted."

"Terrible news," wrote another, stitching his heart to the wrong target. "I hope everything turns well and the sub can be rescued soon. It was a very nice boat. I hope Peter will be free soon."

"Gentelmens," said another, "a real submariner would just silence—and wait 48 hours . . . Dont thrust newspapers to much. At the moment the hole story gives not a clear picture or even makes any sences, The answer at the end will be as allways—simple."

≈

In Denmark, Ole continued pacing. His body seemed so small and thin against the background of fog and sea spray. He wiped his nose and his heart felt funny, like an egg cracking open. The police tried to keep the journalists at a respectful distance. They spoke into their radios, said things like *missing* and *drowned* and *possible* and *impossible.* One policeman muttered to another, as if trying to keep his voice down, as if trying to protect Ole from hearing him, "About 160 centimeters high, 56 kilograms, slim of build, reddish-brown hair and brown eyes." In the depths, the sea monk opened its mouth, took in water.

Ole pictured Kim in his head. He thought of her irrational fear of wriggling creatures—snakes, eels, maggots, even the smallest of worms. But even as she feared them, Ole recalled, she loved them, greeted them with reverence and respect, and appreciated sharing the planet with beings she admitted were miraculous, even if they made her afraid. Ole wished he could erase this day—the weather, the wind, the spray, the police. But he couldn't figure out how to do so without erasing himself, without erasing the love that began to sharpen his sense of loss. With regard to this sensation, Ingrid, Kim's mom, would ask him, 111 days later, on a brutally cold December evening, "Will it never end?"

Ingrid will ask him this from the foyer of her home, which had been overtaken by flowers—bouquets living and dying, petals standing in for condolences. And now, shivering, closing his eyes so tightly against the day, Ole felt that, among the bruise-colored phosphenes, he could see them, sense them—these cumulous bursts of flowers overtaking a hallway. For a brief moment, he recalled the feeling of Kim's arms around his waist, squeezing his ribs, and he tried to place this feeling, and found it four months prior, that April, when he and Kim were traveling in Barcelona. He was driving a moped along the city streets, and Kim sat behind him, her arms wrapping him. They were both wearing helmets and letting the chin straps hang slack, and black vinyl zippered jackets and round sunglasses. They sped past the Generator Hostel, the Club Eurofitness, Mutenroshi Ramen, Gaudí's Casa Milà, the Jardines de Salvador Espriu, and the McDonald's. They were smiling effortless, unbridled smiles. They had pulled to the side of the road, the Carrer de Bonavista, outside the Basar Hongda shop—a variety store selling everything from imported silk scarves to light-up stars and comets, to live tilapia, to dried envelopes of seaweed, to necklaces of dehydrated mushrooms, to lizards that resemble grass blades, to cacti that resemble hummingbirds, to blue-sequined tap shoes, to flowers fake and flowers so real that their syrupy scent overwhelms even that of the fish tanks. Ole and Kim dropped their right legs to the asphalt, bracing the moped as Ole retrieved his cell phone from his jacket pocket and extended his left arm to take their picture.

Ole sneezed now, and opened his eyes now, and the perfect pressure of Kim's arms on his body evaporated, but the flowers were still there in afterimage. "Our home," Ingrid would try to joke, "has been transformed into a florist's shop." And Ole would rub her back, and she would rub Ole's, and their hands would gather static, and their hearts would persist in their breaking.

~

On Sunday, August 13, 2017, Jon Wallace, cofounder and president of PSUBS, weighed in on the chat: "Latest news from copenhagen is that no body found on submarine . . . Newspaper claims officials have determined Madsen intentionally sunk the vessel. I'm not sure how they could determine intent forensically without admission, might just be speculation. Nobody in the sub, but reporter still missing. World attention . . . I have to ask . . . is it possible this is a publicity Stunt? I have read Peter seems to be having a difficult time managing his rocket projects and his latest crowd-funding attempt to refurbish Nautilus was not very successful. I suspect he's got worldwide attention for his projects now."

"Such speculation helps no one," another member responded.

"If this was just a stunt," another member countered, "it will cost hunderthousands [sic] for the raising of the wreck, the investigation, the search mission with ships and helicopters etc. etc. I think we can rule this out."

"I'm not sure I see any rationalization to sinking the sub to hide evidence of a crime in such shallow water that it can be easily reached and recovered," Wallace argued. "So I'm suspect of the authorities theorizing that he scuttled the vessel intentionally for nefarious reasons. However, Peter did some unconventional things, shall we say, with that sub. I recall seeing a video of him swapping out a viewport while it was underwater. I also recall that shortly after first launch when he tried to submerge next to a pier, the aft flooded . . . and the only reason they were able to recover without flooding the sub . . . was because there was a hatch in the fore end that barely sat above water level . . . There was a history of happenstance and perhaps it caught up with him again. Of course none of that explains the missing reporter."

Later that day, Ingrid and Joachim Wall drove across the Øresund Bridge to collect Kim's belongings.

Another PSUBS member answered Wallace, worried that this "Nautilus situation" would once again inspire the authorities to ask questions of the organization. "This is already a media circus, irregardless of what actually happened," he wrote. "This could very well be an event that draws the unwanted attention of the politicians, like we were recently talking about. Do we have any kind of public statement prepared to separate ourselves from this mess? Any reporters contact you Jon?"

Ingrid and Joachim's car shimmied in the wind as they crossed the bridge.

A few days later, on August 21, 2017, Wallace updated the group with fresh information: "Latest news from Denmark is that Madsen has told authorities Kim Wall died onboard his submarine due to an 'accident' and he buried her at sea. I wonder what kind of person it takes to make so many people walk through so much bullshit."

Other members, having done their own independent research, and knowing how a submarine works, immediately questioned Madsen's claims. "He says he was holding the hatch open & it fell on her head when he slipped. I would have thought the hatch would be fully out of the way when he climbed out," said one.

"Not only that, there should be a latch, to lock the hatch open. Especially if it weighs 150lbs! Doesn't take much of a rough sea, to cause an unsecured hatch to slam shut, on what ever, or whoever may be in the way!" said another.

"150 lbs without gas shocks or a counterbalance spring? Skeptical."

"Hi," one heretofore shy member of the PSUBS-MAILIST said, "I've been lurking here for years. I'm a criminal defense attorney and, maybe one day, a submariner. I would just caution folks that one rarely hears the whole story (or an accurate story) through the press. Which is why we are presumed innocent until proven otherwise."

Kim's parents, at that moment, had returned to Trelleborg, Sweden, and Ingrid decided to go back to work as head of information for the municipality, desperately seeking comfort in the performance of mundane tasks. *I can do this,* Ingrid told herself, looking out the window at the ocean, her heart hiccupping when she spotted something small and orange bobbing in the water, her heart sinking when she realized it was only a buoy. Ingrid drove the long way home, along the beautiful coastal road. Not five minutes after she arrived, the press revealed that a dismembered corpse had been discovered along the Copenhagen shoreline by a bicyclist.

~

It didn't take long for the news to hit the PSUBS community. "▮▮▮▮▮ makes an excellent point about getting the full story," one member wrote, referring to the "lurking" criminal defense attorney, then creepily continued, "but what could have possibly required the decapitation and dismemberment of an attractive female passenger and ultimately the sinking of the sub? Someone was definitely having a bad day."

"Only forensic analysis will tell if this body is connected to the 'Madsen Incident,'" another jumped in. "Too early to jump to any conclusions could be an unrelated corpse."

"I have been following the story, and it is pretty clear that something is amiss, but rape and murder allegations are very severe," another said, "and should probably not be thrown around in the absence of conclusive information. Despite the efforts of investigators so far, the only person who likely knows what happened is Peter. Hopefully he will come clean."

As the news developed, and as further details were revealed, the mail list lit up with theories and conjectures, indictments, dismissals, and rumors, some of them plain gross and misogynistic and not worth repeating. Hank Pronk and Scott Waters opted to drop out of the thread early, and Pronk even started a new conversation thread—perhaps as a way to distract from the horror and the ugly gossip—with the subject line "Pressure Test."

Things got petty among the PSUBS group. "I read about this event in the main stream media and thought PSUBS would have the straight scoop. Don't seem to," a member wrote. "So far PSUBS hasn't been dragged into this. Which is a good thing. The public comments are of two kinds. First is sexual innuendo regarding submarines. To be expected since most internet users are morons. Secondly is how unsafe 'homemade' submarines are. Some are and some are not. I would rate Peter's submarines safe. Apparently the good work by you PSUBers hasn't reached to general public. Maybe that is a good thing since most internet users are morons."

"Most Internet users are . . . people like us," another addressed.

"Most social blog writers are morons . . . The quote on morons at psubs is much lower," another defended.

"▆▆, Peter must be very intelligent to have achieved what he has so far," one of Madsen's disciples wrote, "& this is such an ugly mess that it is obviously not planned . . . He looked surprisingly calm when he talked to media after he had just sunk the sub . . . the ladies won't be queuing up for a dive in [another of] his submarine[s] at the moment!"

Even Jon Wallace got a little cavalier. "I'm not sure Madsen has to say anything more," he wrote. "From what I've read, Danish legal scholars believe there is enough circumstantial evidence to convict. I also read that murder convictions rarely result in more than 16 years of incarceration because of their belief in rehabilitation. Anyone want to sign up for a midnight cruise with Madsen when he gets out?"

Two days later, on Wednesday, August 23, 2017, Wallace's message was more sober and included an attachment that has since been scrubbed, but the final clause remains alarming, as if some myth had been busted: "Confirmed. DNA from torso washed up on shore matches that of Kim Wall. Body dismembered and weighted down to sink it. Authorities state they did find dried blood in the submarine that also matched Wall's DNA, so apparently scuttling your submarine to hide evidence isn't foolproof."

Numb and grieving, Kim's parents walked into her childhood bedroom, staring in disbelief at the bags they had brought home from Kim's Copenhagen apartment a few days earlier. They didn't unpack them. What would they do with her things? They were careful to put Kim's laptop—the repository of her work and life as a brilliant journalist—in a safe place.

And on the PSUBS-MAILIST, someone wrote, "Truly bizarre! Were there any clues that he was deranged?"

"Yes," someone answered. "He made submarines!"

## 22

ON FRIDAY, August 11, 2017, at 5:44 p.m., the defense attorney Betina Hald Engmark sat in her car idling at a red light in Odense. She was halfway through her three-hour commute back to Copenhagen from Jutland, where she had filed some legal paperwork at Denmark's Western High Court, a sharp-angled rectangular building atop a grassy knoll, used for criminal cases. From a distance, the courthouse resembles the tip of a giant utility knife surfacing from the earth. Its front entrance lurks behind an imposing and jowly bust of the famed lawyer and former prime minister of Denmark (1853–54) Anders Sandøe Ørsted, who was reviled for his staunch conservatism and was forced to resign from government after being impeached.

As a judge, Ørsted was infamous for his decision in the case of Hans Jonatan. Jonatan was born into slavery on the island of Saint Croix in 1784, which was then a Danish colony, and he labored for the first five years of his life on a sugarcane plantation owned by the Schimmelmann family. When the sugar industry declined, the Schimmelmanns decided to return home to Copenhagen, along with Jonatan. Jonatan was indentured to the family until 1801, when, at age nineteen, he fled the Schimmelmann house and joined the Danish Navy. He played a key role in the Battle of Copenhagen, which saw British ships attack a smaller Danish fleet in the Øresund Strait, with the aim of preventing a Franco-Danish alliance. Though the British were victorious, the Danish prevented Copenhagen from being ransacked. Jonatan's ship was an essential component in the blockade, and Jonatan himself was revealed as a key defender of the city. As such, the Danish crown prince Frederick VI commended Jonatan's efforts, dubbed him a hero, and granted him his freedom from bondage. Such recognition, though, revealed Jonatan's whereabouts to the Schimmelmanns, who immediately had him arrested.

Jonatan's lawyer argued his case before Anders Sandøe Ørsted. Though slavery was still legal in the Danish West Indies in 1801, it had been abolished in Denmark proper nearly ten years prior. Jonatan's lawyer stressed that the Schimmelmanns had no claim on him, and that he was a war hero besides. The Schimmelmanns' lawyer argued that Jonatan was indeed still their "property," as the family had intended to return to Saint Croix to "sell" him there. Anders Sandøe Ørsted considered the case with a fixed scowl and the occasional grumble. In a decision that proved to be an unfortunate landmark on Danish slavery law at the time, he concluded that Jonatan remained the "property" of the Schimmelmanns, and he sentenced the war hero to return to a life in bondage, deporting him to the West Indies, where he could legally be "sold" and "repurchased." Though the details of his escape remain debated, Jonatan subsequently fled to Iceland and eluded the Danish authorities for the remainder of his life.

Engmark, a frequent visitor to the Western High Court, was more than familiar with Ørsted's bust, his defiant scowl now set in stone for posterity. She passed beneath its rotund shadow every time she entered and left the lobby, every time—in her former role as a prosecutor specializing in high-profile economic crime—she earned convictions against those accused of embezzlement, tax evasion, and mandate fraud. She passed beneath Ørsted's gaze each time she visited the court in her former role as police assistant to three different departments. When she did a heel turn, decided to "switch teams" and become a defense attorney, she built her reputation and business on the fact that she knew more than most how the police and prosecutors thought; the nuances of their investigative methodology; how they built their cases. She knew, she often claimed, how to expose the weaknesses of such methodology, and how to dismantle most kinds of criminal cases, including (according to her own press materials) those pertaining to "violence, threats, extortion, deprivation of liberty, and manslaughter," the latter of which Peter Madsen had just been officially charged with.

The press was already beginning to use the language "the case of the missing journalist"; "the submarine case."

On her drive back to Copenhagen, Engmark clutched the steering wheel, and the veins jumped in her hands. She was clad, as was typical, in all black—her leather jacket zipped up, her black sunglasses perched atop her head, holding her black hair back, her black purse slumped on the passenger seat. She wore a thin black bracelet on her right wrist. Her voice was low and hoarse, not because she was sick, but because her voice was always low and hoarse. In its quiet raspiness, her voice seemed to harbor some grave secret, and this quality often inspired people to lean in toward her and listen closely, taking care not to miss any essential information, should it undoubtedly be revealed. Oftentimes (in the public forum at least) her tone of voice is at odds with the content of her speech, the latter of which is typically shrouded in bland legal-speak, and the sorts of platitudes one has come to expect from athletes in the aftermath of a big, tedious win.

In the car, her cell phone rang. The car rocketed forward past the childhood home of Hans Christian Andersen—the house that incubated the author's nightmarish fairy tales of evil doppelgangers and corpse desecration, cannibalism, and self-mutilation. Engmark thought of none of these as she adjusted her sunglasses, as she spat her gum out of the open window, as she sped toward the countryside, as she said, *Hello.*

≈

Engmark publicly acknowledged her role in the case nineteen days later, on August 30, 2017, via an eighteen-second video made available on the Facebook page of DR P4 Radio. Engmark recorded the video in her home office, a low-ceilinged, tight-angled room, painted white, against which the rich rosy finish of her cherrywood desk stood out, appearing more red than brown. On the shelf behind her desk stood an empty glass vase and a stack of thick legal reference books. When sitting at her desk, she often rested her right elbow on the white-painted radiator, and in winter enjoyed its warmth as the heating oil ticked like a metronome and the pipes clanged. But on August 30, by midday, Copenhagen was experiencing lovely weather—a balmy 68 degrees, the fog contributing to a humidity level of 95 percent. Engmark sat in front of her open laptop, the screen further illuminating her face, as if she were standing amid stage lights, her cell phone—the one that originally received Madsen's call—an inert black rectangle on the radiator surface.

"It's not about whether a person is guilty or not," Engmark told DR P4 Radio. "What it's all about is that I have to represent a client. A person. A human. Who is entitled to have a defender. What it's about for me is to carry the person through this process that the person has to go through. I do not have to decide whether a person has been guilty or not."

Engmark was still enjoying some of the electricity that came along with helming such a high-profile case. She had not yet grown weary of the journalistic swarm, reporters following her and surrounding her on her way in and out of courtrooms, restaurants, boutiques, her home. It didn't take long for that weariness to become an almost permanent condition. This case would soon induce in Engmark a seemingly irreversible fatigue.

Soon after taking the case, Engmark was forced to file a restraining order against her neighbor, who in response to her role as Madsen's defender threatened to burn the lawyer's house down. The neighbor, a middle-aged, bent-backed man with spiky hair, ill-fitting clothes, and several broken teeth, often waited for Engmark to step outside, whereupon he would shout such nuggets as, "Betina, you are so cold!" "Betina, you whore!" "Betina, you are lying to the police!" "Betina, I will destroy your life!" and (the final straw) "Betina, I will burn your house!" Engmark, though she declined to comment further on these events, told police that she was losing sleep.

After being held by police for six weeks, the neighbor tried to excuse his behavior to a judge, saying he was being treated for both "psychiatric issues" and hepatitis, after which he simply denied having harassed Engmark, or even knowing her at all. "It's all a lie," he repeated. When the judge suggested that if the man was so bothered by the occupational choices of his neighbor, he should "consider another place to live," the man responded, "No, Judge. Not unless I can move in with you."

~

It's Saturday, August 12, 2017, and families who would have otherwise used this beautiful weekend day as an excuse to picnic by the water, or to take their kayaks or paddleboards or motorboats out into the strait, instead stay home, glued to their televisions, disquieted by the news. Here, into their own backyards, a fresh but as yet indistinct gruesomeness has crept. The female Swedish journalist (whom the press had not yet publicly named) was still missing, and it was announced that Peter Madsen "appears" to have been arrested and charged with murder.

At 10:05 a.m. that Saturday, Madsen—who had been kept awake most of the night by police questioning—was led into the city court in Copenhagen. His mouth was so dry that when he yawned, it made a cracking sound. The police had confiscated as evidence Madsen's olive-green jumpsuit (stained with what a journalist called "indeterminate red plumage"), and so he wore what was he was issued—baggy blue pants and a black T-shirt with *Københavns Fængsler* (*Copenhagen Prisons*) printed in red across the back, bridging his shoulder blades, which in this bright morning courthouse light seemed to be twitching.

Images of Madsen's soiled jumpsuit had begun to circulate in the press and had become a focal point on social media, whereon many debated—sometimes aggressively—the nature of the stains. As is typical of such online arguments, the debates were often mean-spirited and underinformed and pre-developed and rife with false binaries. For some reason, people were desperate to "choose sides"—zealously supporting Madsen or condemning him. Regarding a spatter stain on the jumpsuit's right knee, some said it was obviously blood, while others said it was obviously engine oil.

In the Copenhagen City Court, Madsen sat on a wooden chair, and in the harsh white overhead lights, his face took on a bluish color, looking as if it had grown the membranous skin that forms on the surface of scalded milk. He shifted his jaw, and his dry mouth continued to make those pasty sounds, and those sounds were audible, and likely annoying—as they often are—to the other people in the courtroom, so much so that the interim prosecutor, Louise Nielsen, wrinkled her nose and asked him, "Do you want a glass of water?"

Nielsen's blonde hair was neatly combed, parted on the left. Her face was solemn, bedraggled perhaps by the cumulative weight of her occupational duties—the energy required to prosecute, and the words, gestures, and intonations that attend such performances. She seemed determined to get this initial hearing behind her. It had to do with her body language—her slumped shoulders, the forward angle of her neck. She appeared both eager and resolute, as if she had already known what horrible revelations may lie ahead. As if her experience in law, and with suspects who exhibited the body language Madsen now exhibited—his own alternately slumped and shrugging shoulders hinting at a mean and privileged arrogance, even amid his exhaustion—had trained her to recognize the possibility for such horrors. Both she and Madsen seemed to be enduring this, albeit differently. They seemed to be biting back two distinct versions of impatience.

Madsen smacked his lips a couple more times. "Yes, please," he said, a flat, dull calmness in his voice—a voice distant and detached. He was given a glass of water, and he sipped. By his side was his interim defender, Finn Bachmann, a gnarled and burly man, his head shaved to futilely distract from his natural baldness, his smile so tight it seemed as if he were growling through his exposed bottom teeth, the collar of his white button-down loose, his black sport coat bunching at his armpits. Though Bachmann prided himself on his personability and directness, his operation at the "highest professional level," and his willingness to offer a client "realistic expectations" to the outcome of their case, Madsen wasn't pleased with his counsel, offering Bachmann only sidelong and dismissive glances. He wondered aloud as to the whereabouts of his requested attorney, Betina Hald Engmark.

The judge, Kari Sørensen, assured him that Engmark would be arriving later that day, well before the actual constitutional hearing began.

"I really want to talk to her," Madsen said, a fresh waver colonizing the calmness of his voice.

Sørensen sighed like a parent resigned to assuaging yet another irritable child. "And you will probably get to that," she told Madsen.

Murmurs circulated among the journalists and photographers who had packed the courtroom. As there weren't enough seats to accommodate all of them, many stood leaning against the walls, jotting onto notepads, taking notes on their cell phones, huge cameras necklaced around their heads. They had traveled to Copenhagen from all over Denmark and Sweden. One of Madsen's mistresses arrived early enough to secure a chair and, wishing to protect herself from the paparazzi, gave them conflicting reports as to the nature of her identity. Most of the papers would refer to her as Madsen's "girlfriend." She arrived at the courthouse carrying an overstuffed black sports bag, which she tried to get to Madsen, but was made to turn over to investigators (who also packed the courtroom). The contents included a change of clothes and toiletries, which she had brought at the suggestion of Engmark, who—even at this early stage in the proceedings—felt that Madsen was going to remain in custody for some time.

It was warm in the courtroom, the outside temperature already in the mid-60s, the sky cloudy, and the humidity increasing. By 6:00 p.m., it would be raining. The deodorant at Bachmann's underarms was pasty, and this, paired with his snug sport coat, limited his movements, so his gesturing (and, in turn, his performance as a defender) felt stiff, robotic, phoned in. As Louise Nielsen intoned, "Peter Madsen is charged . . . ," Bachmann's nostrils twitched, as if he had felt something fluttering in one of them—a wayward hair or booger—but decorum kept him from addressing the itch with his finger.

". . . with having killed . . . ," Nielsen said, and Bachmann exhaled forcefully through his nose, as if trying to expel the little obstruction with breath.

". . . Kim Isabel Fredrika Wall . . . ," Nielsen said. Madsen sat with his hands folded in his lap, his eyes rinsing over the members of the press at the rear of the courtroom, his mouth smiling.

". . . in an unknown place . . . ," Nielsen said, and Madsen's smile grew, as if he were picturing the place—*Nautilus*—with which he was so intimate; the place with which he would never be intimate again, and his smile was the smile of someone harboring a secret, cut with an odd mournfulness. It was the expression of someone who, in middle age, was ruminating on their bygone days of high school football field heroics, the sort of which made them feel superior to everyone else in the room, however unrecoverable such so-called heroism may be. In short, it was a pathetic expression, though Madsen, of course, was oblivious to its pathetic quality, a quality that, of course, depended on such obliviousness.

". . . and with an unknown method after 19:00," Nielsen concluded.

Murmurs again circulated among the members of the press, and their voices increased in volume until they were murmurs no longer, and Madsen sat, his hands in his lap, and he seemed satisfied with their reactions, their enthusiasm. Judge Sørensen did not share Madsen's apparent satisfaction, though, and ordered the courtroom cleared of its audience, and the doors closed. The members of the press exploded in protest, shaking their pens, their cameras, their fists, shouting about the unfairness of it all, about how they had the right—no, the *duty*— to remain on-site and report the minutiae of the hearing to the thirsty public.

Nielsen tried to side with Judge Sørensen on the matter. Bachmann remained silent, as if he too were rapt by the chaos and could do little but spectate, wiping his nose, scratching at his armpits. Judge Sørensen tried to maintain order, to speak over the howls of the press, who were busy being shuttled toward the exit by security, turning back over their shoulders and spitting their complaints. Madsen appreciated the circus, leaning back in his chair, folding and unfolding his arms, a smug grin spreading across his face, his eyes bearing a near-beatific intensity, his forehead wrinkling. He was nodding, as if in agreement with the journalists, as if he was on their team and they on his, Rocket Madsen once again seeking (and perhaps inventing) the team of disciples he so desperately needed. The more the journalists protested, the larger Madsen's grin, the more vigorous and confident his nodding.

Into this melee, Engmark speed-walked, "rush[ing]," as one newspaper would later report, "into the courtroom last minute." She was huffing, a few strands of her black hair having loosed from its tie, her sunglasses atop her head, black purse under one arm, black briefcase clutched in her hand. She broke through a barrier of journalists, thrusting their red-topped microphones at her face as if roses.

She had been at the police station, which was only two kilometers away, but it took ten minutes by car to arrive here. She'd sat rubbing her eyes in the backseat, shifting nervously, as her driver navigated the traffic. At a stoplight, they idled at the intersection that housed the Hotel Absalon (named for that ruthless twelfth-century Danish politician), which hadn't yet turned off its blood-red neon facade lighting, though it was daytime. Engmark shifted in the backseat and looked at her black-banded wristwatch. Her meeting at the police station had taken longer than expected as sometime during the previous night the police had confronted Madsen about the inaccuracy of his story—that he had docked *Nautilus* and had dropped Kim Wall off at 22:30 outside the Halvandet restaurant.

Apparently, they had shared with him that the CCTV camera that surveilled the area showed no such evidence. They had played the video, which had a time stamp at the bottom, and there were, of course, no images of Madsen, and no images of Kim Wall—not in the vicinity of 22:30, and not before, and not after. On the screen, other residents of the city milled about, backlit and front-lit, their shirtsleeves ruffled and billowy, the fog or mist accumulating, phantasmal, in the air over their heads. Apparently, Madsen had intently watched this video—the people with the wind-puffed sleeves, the ghostly mist, or the ghostly fog, as if scrutinizing the screen for himself and for Kim, or another version of himself and Kim, conjured versions that confirmed his fabricated story. As if hope and wish had had any place in this police station, at this hour, in the aftermath of such an atrocious and monumental event. As if hope and wish had the power to undo murder; had the power to render that which occurred at depth to the realm of dream—watery and muddied, distended and dizzy.

Finding no such versions of himself and Kim, Madsen had dropped his gaze to his hands, and he watched them rubbing one another in the bright glare of the police station lights, and for a moment they seemed as if they belonged to someone else—someone who had no business being here, in the cruel inspecting hardness and brightness of an interrogation room on dry land. Then he had lifted his gaze to the officers, smirked and sighed, and, in a revision that would—the next day, Sunday, August 13, 2017—inspire various sub-headlines in various newspaper articles (all bearing some variation on "Rocket Madsen Changes His Mind"), pivoted to a different narrative, one he had perhaps rehearsed with Engmark, perhaps not.

If one were to judge by Engmark's tardiness to the courthouse that Saturday, and her breathless entry, one could assume that Madsen had gone rogue, and that she was as surprised as anyone at the revelation. Madsen had told the police that he made a mistake, he was scared, and exhausted, and distraught, and confused. He had the slick of the deep on him like a hangover. He had told them, yes, there wouldn't be any evidence of his dropping Kim off that night at Halvandet. That's because, he had told them, there had been an accident. A terrible, terrible accident.

"If water is clear," asked the staff writers at *Scientific American*, "then why can't we see through it clearly? [W]hat happens if you try to look through it to see the world on the other side . . . ? It looks a little distorted, maybe a little fuzzier and uneven . . . Similar to when you try to run in a swimming pool, when light tries to move through water or glass it gets slowed down. When light is slowed down, it either bounces off the material or is bent as it passes through. We can see these changes in light, which indicates to us that something is there."

That night in the police station, Madsen had been simultaneously looking at the world on the other side of the water and occupying space in it. He was slowed down and bent, uneven, but there. He had decided to try out another lie, test the power of his fuzziness. Kim Wall did indeed die aboard *Nautilus,* Madsen had told the police. He had spouted off the story about her insistence to climb the submersible's ladder in order to snap a few photos, in spite of his own warnings against doing so, especially while the vessel was surfing the chop; about how, as the sub bumped up against a swollen wave, the hatch door swung closed, hit Kim on the head, and knocked her down the ladder. "She is dead as a result of [this] accident," he had told the police, then sighed, and added, "I buried her at sea."

Engmark now pushed her way through the journalists, Madsen still nodding approvingly at their protests against the clearing of the courtroom and the closing of the doors. Nielsen was still defending her request to both the press and the judge, reiterating that the "relatives of Kim Wall may be violated by the information that will be revealed at the hearing," to which several journalists objected, claiming their presence was necessary to promote "legal certainty" and to respond to the "public interest in the case." As Nielsen sparred with the press, and as Judge Sørensen pleaded for decorum, Engmark rushed to Madsen's side, gliding over the floor as if it were ice. She bent to his ear and whispered. He lifted his face and whispered back. The different forces and temperatures and aromas of their breaths twined in the air between them. The overhead lights glinted off her earrings. The sleeves of her bomber jacket swished. Madsen sniffed. Amid the whines of the journalists, Engmark took to the center of the floor, opened her mouth, and inhaled. The courtroom went quiet.

"My client," she said, her low raspy voice sounding grave but electrified, "does not mind giving his explanation while the press is present. In fact, he really wants it."

Nielsen shook her head, having prized the potential wants of Kim Wall's family over Madsen's.

"This case is of such a nature that the public has an interest in knowing what has happened," Engmark continued, and the members of the press nodded, as if vindicated.

Nielsen rehashed her arguments, and the journalists groaned, and Engmark raised her voice and spoke over them, rehashing her own. "My client would like to give an explanation. He is cooperative and would very much like to give an explanation in front of the press."

"Very much like that!" Madsen himself broke in, and Engmark's shoulders dropped at the intrusion of his voice, and Nielsen looked at Judge Sørensen, and some spell lost its power to bewitch, and Sørensen confirmed that the courtroom was to be cleared immediately, and the doors, indeed, closed.

As the dejected journalists and onlookers rose from their chairs, Madsen caught the eye of his mistress. He raised his eyebrows and brought his fingertips to his mouth, kissed them again and again, then threw the kisses to her. If she caught them, nothing in her demeanor indicated it. But Madsen kept kissing. The sounds of his lips on his fingers were obscene. He sounded as if the air nozzle of a diving suit were malfunctioning. The sight of it was obscene as well, his thick fingers, chapped knuckles coming to his wan lips. If the din of the onlookers, as they made their way to the courtroom doors, hadn't drowned out the sounds of Madsen's mouth, surely said sounds would have aroused such an intense and physically manifested annoyance—the sort that began vaguely in the perineum and traveled up the spine, like fear, branching out at the base of the neck and creeping into both ears—that they would have been compelled to grind their teeth before screaming at him to stop it. To just stop it, for the love of god.

Neuroscientists tell us that when we touch ourselves in intimate but nonsexual ways, our brains release oxytocin—the "bonding hormone"—which further compels the discharge of other happy-making hormones such as dopamine and serotonin, as well as stress relievers like cortisol and norepinephrine. As Madsen kissed his own fingers there in the courtroom, his heart rate dropped, and his blood pressure dropped, and he alleviated his anxiety, staved off depressing thoughts, and, subtly, boosted his own immune system, and relieved—however slightly—any pain he may have been feeling. Perhaps, as the neuroscientists would have us believe, this resulted in Madsen feeling more generous than usual, more empathetic, nurturing, collaborative, and grateful. And in this way—like Alexander the Great before him—an atrocious man was able to calm his own heart as he prepared to confront the atrocities he perpetrated, perhaps even believing that, in abating his own anxiety about the trial, he would shrug off a measure of accountability as well.

Microphones hanging flaccidly at their sides, the journalists exited the courtroom, and the remainder of the onlookers followed. And the light in the courtroom appeared newly sallow and unnatural, and if one were so inclined to stretch the imagination, one may imagine that the entire building was not on dry land at all but submerged within some great body of water.

# 23

**O**UTSIDE OF COPENHAGEN, and in middle America, a tree falls in the woods.

Two hundred miles east of the house where the Clutter family was murdered (the crime immortalized in Truman Capote's *In Cold Blood*), and an hour-and-a-half drive from the geographical center of the contiguous forty-eight U.S. states, so far from any ocean, the amateur submersible builder Scott Waters logs on to the PSUBS email chat board and weighs in on the thread everyone is talking about. It's the middle of the night, and quiet, and the computer screen blues his face as he types, heart racing, "My electrical engineer is in Denmark and knows [Peter Madsen] well and is currently there. The submarine sunk in very shallow water (like 7 meters)."

Outside Waters's window, the Kansas camel crickets—the quietest crickets in the world, with no sound-producing organs—burrow into his drainage pipes. It isn't them out there in the fields that he hears chirping. His face is blue. He is exhausted but wired. He perpetuates the same misinformation—based on Madsen's initial claims—that other p-subbers were perpetuating at the time. He types, "There was a female news reporter on board. After they were rescued, he took her home then she was reported missing. Everything is currently under investigation."

How did it come to this? Waters recalled a so-called more innocent time, four years earlier, when he himself felt so much younger, as he prepared to dive in his first home-built submersible.

~

The Salina, Kansas, native did not change his last name to suit his obsession. He was born this way. Twenty-seven years later, it is September 22, 2013, a hot, humid day in Salina, the heat peaking at 84 degrees at 2:53 p.m. It's a wet heat. A sticky heat. In videos of Waters being interviewed by local news sources, he's often wiping the sweat from his eyes. The cicadas sound like buzz saws. Dandelion fluff and smears of dirt cake the sweat at Waters's neck. He is so far from any ocean, but he is undeterred and has been undeterred for most of his life.

"I watched [the Disney movie version of *20,000 Leagues Under the Sea*] over and over, and I got blueprints for a submarine back in the first grade," he says. "I actually knew back then how a submarine was built."

Waters was born into a family that runs the local Salina hardware store. He grew up crawling its aisles and playing with the hoses and lightbulbs, the nozzles and nails. He learned from an early age what sharpness was. He learned which tools would smash things most efficiently.

"I was a . . . curious kid," Waters says. "I made blueprints and drawings of how I would design my vehicles and had a good understanding of how they worked at a young age from books I would read. I would also get any toys that had buttons and knobs and fashion them into some sort of imaginary vehicle and dream of going to places no one has been to before . . . I was bullied a lot."

He cites his time as a Boy Scout and his earning of an Eagle Scout rank for braiding a seriousness into his curiosity; a drive to get things done, make things happen. He christened his sub *Trustworthy*, after the first law of the Boy Scouts ("A Scout is trustworthy, loyal, helpful, friendly, courteous, kind, obedient, cheerful, thrifty, brave, clean, and reverent"). "But if I had known how long it was going to take," Waters says, "I would have named it *Persistence*.

"When I started my first business [as a senior in high school], which was a party and wedding event rental business, I made enough money to build my first submarine," he says. While most of his classmates at Salina South High donned their green, gold, and white cougar apparel and mingled at the football games, Waters tinkered in his garage and peddled rental tuxedos.

"[Later], I dropped out of college after five semesters because I didn't like it... [To build my sub] I had to [independently] learn and master all the skills like welding, electronics, fiberglass, and machining... I was extremely aggressive."

In one 2013 video, Waters wears a black T-shirt, snug at his armpits, and khaki shorts that extend past his knees. He is wiry and strong. His outer arms are sunburnt; his inner arms, milk white. His hair is dark and shaggy, his eyes droopy but alert. His stubble, a couple days old. He is flanked by Milford Lake, a dollop of water in the middle of the corn- and wheatfields. From its banks, one can spot the tops of distant decaying silos, caved in like stamped-out cigarette butts. One can hear the moans of the area's livestock. The lake's inflow source is the Republican River. Most folks come here to fish for bass and walleye. It has a maximum depth of sixty-five feet.

"Thousands of hours," Waters says. When he lifts *Trustworthy*'s hatch, his longtime friend and crewmate, Carl Boyer, pops from inside like an emaciated jack-in-the-box. Boyer is even more wiry, his cheekbones pronounced. It seems as if his skin is too tight for his skull. He sports a pencil moustache and shoulder-length dirty-blond hair. Waters holds the hatch lid aloft for Boyer, and though Boyer's emergence seems like an act of whimsy, there's no trace of whimsy on his face. Boyer stares into the distance, scowling at the sun. His yellow T-shirt is the exact shade of the submarine, and he rests his skinny arms on the hatch rim.

Posed like this, both men boast about the specs of the sub as if they're talking only to one another: that it was built piece by piece over the course of five years (begun when Waters was only twenty-two); that it's fourteen feet long, six feet tall, and four feet wide (inside, its dimensions are less than that of a full-size mattress); that it's steel-plated; that it weighs forty-five hundred pounds, about one-third of the weight of a male elephant.

"Built it from scratch," Waters says.

Waters and Boyer go back and forth naming the sub's features as if one-upping the other: boiler-grade steel, high-pressure air system, enforced framing, enforced bumpers, primo acrylic windows, max. depth of 350 feet...

The thing can travel at only 4.5 miles per hour, but who wants to go faster when bobbing among the walleye and the wild celery in the largest man-made lake in Kansas, the construction of which demanded the relocation of three cemeteries and the razing, burning, and burying of two towns, plus the sturdiest grain elevator in the state? The sub slugs along the muck amid the flooded remains of these towns—homes and restaurants and railroad tracks, stockyards and schools, the famous cottonwood tree under which many of the area's residents were conceived, the wreaths of dried flowers that once adorned graves.

The yellow bean of a sub roves amid the splinters and the inscrutable evidence of centuries-old floods, remembered now only by the ghosts and the stratified mud—the flood of 1781 that killed many of the Sioux, Cherokee, and Iroquois Nations; the flood of 1849 that saw a great herd of buffalo drowned, and when the water level fell, the treetops along the area rivers were overspread with their bodies, and a frost came thereafter, preserving the carcasses, and all winter long, the residents would scale the trees with their knives and carve away enough meat to feed their families for weeks; the flood of 1869 that saw parents scale these same trees and tie their infants up into red blankets and cinch them to the highest branches as the waters rose, and those parents descended and drowned, and the waters again fell, and when a rescue team was finally able to enter, they found only trees adorned with these horrible blankets, the infants who survived screaming inside them, these writhing, wailing oversize apples.

Amid these swamped artifacts and stories, Waters and Boyer can bob—Waters as the pilot and Boyer as first mate.

"The pilot drives it like an airplane," Waters says. "I'm also working on my pilot's license, so I think I'm going to build an airplane [too]."

They back the sub into the lake with a trailer, as other folks do with skiffs and motorboats. If there's an emergency before the sub dives beneath the surface, "I can use a slate or sign language if anything goes wrong," Waters says, and if something goes wrong when fully submerged, "I have a full scuba system inside," though he acknowledges it may be difficult to escape the boiler-grade hull while underwater.

It's nearing 3:00 p.m. Waters pats the side of the sub. "There's about sixty of these in the whole world," he says, meaning personal home-built submersibles, "but [this is] the only one in Kansas."

Boyer nods along, and Waters finally gives him his props, calling Boyer a crackerjack electrician who designed *Trustworthy*'s entire electrical system. "Yep," Boyer says, also patting the sub. "It's exciting. I mean, we're building a sub in *Kansas*."

A Kansas wind ribbons the lake's surface. Like confetti, quaking aspen leaves burst into the air before settling on the water, and on the sub.

"It was a fight and a struggle," Waters says, and Boyer nods, "but if something feels impossible, I keep punching at it until it's done."

A fisherman in yellow waders walks by holding what must be a sixty-pound fish in his arms—one of the famed Milford Lake blue-cats, an endangered catfish once thought to have gone extinct, thought to have been a sea monster, a relic of prehistory, a predictor of disaster. In Shinto mythology, for instance, the giant blue catfish Namazu lived deep beneath the seabed, and its swimming was the cause of all earthquakes. In 1933, inspired by the stories of Namazu, preeminent Japanese seismologists Shinkishi Hatai and Noboru Abe observed that catfish in "natural" aquaria displayed increased agitation about six hours before significant earthquakes occurred and could be registered on their recording apparatus. Blue catfish—at least as we humans perceive them—are lazy, placid, and unresponsive creatures, so this display of intense restlessness proved uncanny and shocking. Hatai and Abe determined that the blue-cats were able to predict earthquakes with over 80 percent accuracy.

The Milford Lake blue-cat slumps in the fisherman's arms like some legless Rottweiler. It has a beautiful full face. These fish are not supposed to be kept but returned to the water after being caught (and photographed, of course). The angler walks past and since his waders are yellow, when he crosses *Trustworthy,* in a trick of the sunlight, he seems to disappear, and the giant fish appears to be floating midair—some slack bride being carried over the threshold by a phantom.

But Waters and Boyer don't seem to be seeing this, and if they are, they pay the vision no mind. They both have a hand on the sub, as if plugging themselves into some life support system, recharging. They squint into the distance, toward the crumbling silos. They both nod, fingers on *Trustworthy*. Waters stares over the water and, to no one or nothing, it seems, says, "I just can't stop."

≈

It's three years later, 2016, and Waters is building a more ambitious second sub, of which he reports, in understated monotone, "When we finish this, we will be the deepest-diving private individuals in the world."

Back in 2013, after he finished and launched *Trustworthy*, he "kinda got almost depressed," he tells me, "because it's like I've been working on this for so long. Now what? . . . But I always kept thinking: What if I build a deeper-diving submarine? And my original thing was I thought I'll build it from scratch. I realized that was not something I was able to do or afford. So, I [had] to figure out how to buy a used deep-sea submarine. I started in Russia . . . and I got the opportunity to get *Pisces VI*." The asking price was $500,000, and Waters somehow negotiated it down to $30,000.

By this point, Waters's family was running a chain of True Value hardware stores, and Waters himself served as the CEO of the family business for three years. "And I realized I didn't like that very much," he says.

He stands in his brightly lit workshop—tucked away in the cornfields—a tall orange ladder open beside him, his sub-in-progress, *Pisces VI*, looming behind. "There are only five submarines in the world that can dive as deep as this one," Waters says. "It's like building a spacecraft."

At this stage, *Pisces VI* looks like a brassy meteor, an archaic bomb that has been unearthed from the desert; Tatooine from space. He works into the night, and the night is cool, but he's still wearing a T-shirt—a red one—and he's holding his own arms as if he's cold. Waters says the cavernous workshop was custom built to accommodate *Pisces VI*. "We specially formed the concrete for it—seven-inch-thick slabs so it could handle the weight."

On his T-shirt, arced over the left breast, is the emblem of USS

*Nautilus* (SSN-571), the world's first operational nuclear-powered submarine—the same one on which Admiral Hyman G. Rickover, the disturbing muse of Commander Scott Waddle, was once stationed. In 1958, USS *Nautilus* became the first sub to ever complete a fully submerged transit of the North Pole, beneath the Arctic ice sheet. This particular mission, called Operation Sunshine, was initiated in response to the nuclear threat posed by the Soviet satellite Sputnik and involved the testing of the vessel's ability to launch ballistic missiles.

This nuclear *Nautilus* saw its first launch in the Thames River and was sponsored by Mamie Eisenhower. Oddly enough, *Nautilus*'s ship patch was designed by Disney. Two years after her launch, this nuclear *Nautilus* logged her sixty thousandth nautical mile, matching the distance traveled by her fictional namesake in Jules Verne's *Twenty Thousand Leagues Under the Seas*.

She was decommissioned in 1980 after it was determined that the unusually loud noise she emitted would make the sub vulnerable to enemy sonar detection. *Nautilus* was declared a "National Nuclear Landmark" by the American Nuclear Society and put to pasture in Connecticut's Submarine Force Library and Museum, where the vessel sees about 250,000 tourist visits per year.

Waters's Kansas workshop is a massive khaki outbuilding on his family's property, fronted with trailer hitches, a navy pickup truck, a dented silver minivan, all parked on the greenest grass I've ever seen. Standing inside, Waters says, grinning, of his fate of having been born in the middle of the country, "I try to look positively at it. I'm close to either ocean." He says this as if it's a catchphrase. It sounds well-practiced, and I'm sure he repeats it to most any interviewer. He picks up an acrylic window lens, holds it reverently to his face. "Just thinking of all the different scientists that looked through this . . . ," he muses. "I mean, this whole entire project is to build a low-cost platform for science, film, for tourism. Basically, if you say, *I wanna go down to six thousand feet and take a look at what's down there,* and you have the money to do it, you're gonna spend $40,000, at least. That's just to go down and touch it and come back."

Waters yawns; he looks as if he hasn't slept in days. He will later tell me, "I have dedicated a lot of my time to [building and diving in submarines], which means I spend less time doing other things. I never watch TV, and my friends and family know I spend all my time working."

One of Waters's volunteers, a local kid who looks about fifteen, crawls like a spider on top of *Pisces,* constellations of freckles on his cheeks, an arm-length wrench in his hand. "It's mind-blowing," the kid says in a slurred drawl, clinging to the sphere. "When he's got this sub, and he's asking for people to help volunteer [to] take it apart, I am *there.*"

Another volunteer with gel-stiffened blond hair lowers himself into *Pisces.* "It is tight," he says. "It is amazing how small this whole thing is. It's six feet wide from side to side." Waters's crew is working on removing most of the old equipment. Once finished, Waters hopes to get *Pisces* down to eight thousand feet, "possibly even lower." He feels he can finish it in about four years.

It's easy to lose track of time beneath the workshop's white fluorescent lights, behind safety goggles or a welding mask, or enwombed in the sphere itself, contorting one's body to the body of the sub as if a snail, winding itself into its spiral shell. It's easy to exclaim in here, to be immune to the holiness of one's own excited voice commingling with others, shouting while building a submarine together, such tiny eccentric stitches in the tapestry of these great plains. It's easy to work through the night, and to emerge before dawn when the light is dim, milky blue.

Waters looks at the sky and points out the stars as if he's pining for them. For him, too, the attraction to deep water and to outer space overlap. "I have loved submarines *and* spacecraft since before I can remember," he says. "I've been inspired by space exploration and ocean exploration, getting in gas balloons and going super-high in the atmosphere . . . Really just going places where humans are not really supposed to go has always interested me."

The Big Dipper hangs over a cornfield, and Waters exhales. "The most profound moment is the first time you go very deep," he continues. "I would compare it to the feeling every astronaut feels the first time they see the curvature of the earth and their perception changes about humans on earth." And, indeed, though he doesn't yet know it, Waters will in just over a year briefly serve on the UN Space Advisory Council in the human spaceflight division, and in two years he will briefly serve as a private contractor for the United Arab Emirates space agency. "The more I work in the space industry," he will say, "the more I see the similarities [between rockets and submarines]."

On Waters's dirt driveway, some of his crew chain-smoke cigarettes and toe the deepest of the ruts, the backlit corn rattling in the wind. A rooster calls. The men stretch their bodies. They look like weather vanes. They look like a campfire crackling, the orange cherries of their cigarettes wending in the sky. Tawny chickens scramble up the driveway. Something has spooked them, and they find new ground from which to peck on the back side of the workshop in which *Pisces* sits on its platform, some queenly orb preparing to predict the future—an earthquake, a typhoon, a drowning, a rescue mission; the death of all of these chickens, and some of these men.

≈

It's September 2022, and Scott Waters has just returned to his apartment on Tenerife, Spain, in the Canary Islands, where he has been living for the past two years. He's exhausted, having been aboard *Pisces VI* off the coast of Lebanon—about six thousand kilometers from Tenerife, searching for a sunken refugee boat, and the bodies of the drowned, missing since April. The boat had been carrying a group of people from Lebanon, Palestine, and Syria who were trying to make their way to Italy when it sank beneath the surface about five kilometers off the port of Tripoli.

Accounts pertaining to the reasons behind the sinking vary, but according to the Associated Press, the boat went down "following a confrontation with the Lebanese navy." Thirty-two people went down with the ship, and when Waters and his scant crew (*Pisces VI* can accommodate only three) began their search for the remains, it was 88 degrees and sunny, and the sea breeze felt as if belched from a furnace. Before the search began, *Pisces* was towed to the dock at the navy base in Tripoli and brought up with a crane. Waters, clad in a tight navy shirt and khaki shorts, climbed atop the sub to inspect it, his calf muscles jumping as two naval officers in beige fatigues and black berets stood guard beneath him, their fingers on the triggers of leg-long guns.

Waters has built a private company around the submarine he cobbled together in his Kansas garage, and, like Frankenstein's monster, it has led its creator into unforeseen and harrowing places. Waters still can't believe it.

"As far as the Lebanon mission . . . um . . . basically . . . um . . . ," he begins telling me, back on Tenerife, before pausing for a long time, as if parsing through the ornaments of the trauma. He won't return to the details of the mission until later in the conversation. He swallows hard and refocuses his attention on the vessel itself—his beloved and infernal submarine, the mechanism that for better or for worse changed his life. "Almost all the submarines that are like us—they're called DSVs [deep submergence vehicles], and it's any submarine that goes below two thousand meters. Almost all of them are owned by governments . . . even if you had a lot of money, you couldn't pay to go on them. An exception to that was the *Mir* submarines—the Russian subs"—one of which, in 1998, went on a futile search for two tons of gold and three tons of assuredly ruined opium that were supposedly in the hold of a sunken Japanese submarine on which fourteen people died during World War II—"and if you've seen the movie *Titanic*, that was how they did that [documented the undersea wreckage], because back then, there was no access to the ocean unless you hired a government, and Russia was the one willing to do it. So, I came up with this business idea . . ."

He's wearing a navy-blue cap, and his brown hair is now long and tied up, strands of it dancing from the sides of the brim. He often puts his hand to his mouth as he speaks, stretching his lips, running his fingers over his stubble, as if carefully considering something I can't see—as if both aghast and bored with being aghast. He has dark circles under his eyes, and one gets the impression he hasn't been sleeping well. There's something oddly indistinct about him, as if he's fading away, ready to become static, dispersed into the sea.

The weather here is so comfortable as to be invisible. As Waters is fond of saying, "It's never very hot and never very cold," and so it's nothing really. It negates itself. If the wind isn't blowing, the body doesn't register anything resembling weather, and this feels destabilizing—some sensory deprivation that makes one long for a tether, long to feel something assert itself on the arms or legs or hair. One feels like they're floating, adrift. Down the street, in the town of Candelaria, the basilica bells ring out over the sea from the crooked white belfry, and wraithlike beachcombers hunch over the sand, searching for anything beautiful, anything of value. It was here, on this beach, that the Virgin of Candelaria appeared miraculously in 1392. It is here, on the adjacent plaza, that visitors from many mainlands down shots of ron miel (honey rum) topped with whipped cream and cinnamon at the Amore Mio bar flanked by bronze statues of the nine indigenous kings who ruled here in pre-Hispanic times.

Waters holds his hand to his mouth before dropping it. "The only missions that we do are ones that are for helping people," he says, "that are not conflict, or could be harming people . . . [But] this mission in Lebanon was an interesting one because . . . it's not harming anybody, and it's only helping people, even though there's . . . there's . . . there's thirty-two people deceased, but we were however providing some peace for the families that have, you know, that have lost family members."

Waters explains that he and *Pisces VI* were commissioned by a "relief fund" NGO known for their humanitarian work in the Middle East to try to locate and salvage the refugee boat. "They were looking for a piece of equipment that could do this," Waters says, "and we won the bid," which was reportedly $450,000. He bites his lip, as if he's scratching at the implications of *won*. The bells stop ringing, and there's a techno beat playing in the distance, and the buildings are painted so brightly, rich blues and pinks, and oranges and purples. His eyes are bloodshot, and in the wake of the *Pisces VI* mission, Reuters, reporting from Tripoli, is busy publishing an article with the headline "Families of Migrant Boat Victims Find No Closure in Crisis-Ridden Lebanon," even as they were allowed to witness, from the deck of a nearby Navy ship, Waters's submarine begin to search for the bodies of their loved ones. It was reported that the sub, "amid the still blue bottom of the Mediterranean Sea," looked so small.

*Pisces VI* is now being towed across the Mediterranean from the Lebanese coast. It will be a few days before it arrives at its spot at the marina in Radazul, which also houses a chichi boating club, some fancy beach bungalows, and the best-rated Italian joint on Tenerife. Waters lets out a long, shuddering sigh and shakes his head. Though his sub is out to sea and so far away, based on Waters's glassy stare into the middle distance offshore, it seems as if, in his head, he's still on board. When he speaks again, his voice bears a mournfulness, and everything he says sounds soft.

"We knew where the boat had originally sunk," Waters says, "but that doesn't mean that's where it is. So we had done the first dive about one kilometer away from the actual wreck site, moving up the wreck site—from down-current to up-current—to locate the debris field. We found it quite quickly." He keeps his eyes on the water.

"This mission was unique," Waters says, ". . . it was such a massive loss of life . . . This is very much outside of what we normally like to do . . . But, yeah, so we did this." His mouth sounds dry. His hands are kneading one another. "The first dive," he continues, "we located a body . . . The first body . . . We decided to try to bring the body up, though the body was almost completely deteriorated. Um, pretty much just the clothing had made it to the surface because the body itself was kind of like dust."

And the dust that was this body burst into the sea as if from a popped balloon and whirled over the coral beds, the sea having rendered this person to ash, doing the job of the crematorium's fire. "As it was hoisted to the surface," Reuters seconded, "the body disintegrated piece by piece . . . leaving only clothes in the grip of the submarine's robotic arm."

Tom Zreika, a member of the NGO that hired *Pisces VI* and a former refugee himself, tagged along with Waters on the mission and revealed a detail Waters can't bring himself to mention. "There was a woman down there," Zreika said, "stuck halfway out a window holding her baby . . . That's the one that broke everybody's . . . ," and he trailed off without having to say, *heart*.

"Ugh," Waters utters, and exhales hard. It seems as if he's struggling for air. "It just, just, it, it, fell apart, basically . . . We then made the decision that the next bodies we encounter, we will document their location, but then continue to locate the wreck rather than, um, stop the dive to bring up, uh, stuff from the bottom." And I wonder what makes Waters—after intoning the word *body* so many times already— catch himself, as if he can't say it again, as if he's maxed himself out, and replace it with the generic *stuff.*

Referencing the woman and her baby interred at the seafloor, Zreika had said, "It reminded me how my mother would have been holding me."

"So, we, we did that," Waters says, his voice cracking at *that.* "The next dive, we found the wreck. And we spent about an hour and a half documenting it, just with, with film. And, you know, uh, uh, this is what it looks like."

And I'm not sure what he's envisioning, and he certainly doesn't offer to show me any of the footage, if he even has it in his possession. But we know the boat was in pieces and that scraps of clothing waved from the splinters. And we know about that woman and baby and that there were more corpses down there, scattered about the wreck site, and we know they were barely holding together and that with one too-fast whip of the current, they wouldn't bear the shapes of bodies anymore. We know that, for many of the families of the deceased, there has been and will be no peace. And we know that, because Reuters tells us so: "Distraught families of the missing hoped the submarine would reveal the fate of their loved ones and how the boat came to sink. They now fear the bodies will remain submerged for good, along with potential evidence."

"They kinda wanted to do an investigation to figure out what happened," Waters says, "because there was hearsay that the navy actually had rammed this boat. That turned out not to be true at all." Though it was reported that a Lebanese Navy ship "intercepted" the migrants' boat, Waters claims the wreck itself "showed no clear signs of damage from [the] navy boat."

Many family members of the deceased distrust this analysis, and few of them, according to Reuters, "expect it to come to a just conclusion, underlining deep popular mistrust in Lebanon's judiciary and state institutions . . . Some families of the victims have filed criminal charges against the navy officer commanding the ship that intercepted the migrant boat." There were, in fact, originally nearly eighty refugees on board the boat, and many of the survivors indeed "said they had been rammed."

"It's a mockery in every sense of the word," said a man whose son drowned, "that any self-respecting institution would be the perpetrator accused of killing people and also oversee the probe."

Off the Candelaria coast, there are waves in the water, but still, I can feel no wind. Something's not right about this. Waters kneads his own hands. "And, so, anyway, we decided . . . it just wasn't feasible [to bring the boat up]," he says. "It didn't make sense. And so, we decided to leave the wreck down there."

According to Reuters reports, the mission was called off by the Lebanese military, which cited "possible security risks . . . An attempt by a lawyer for the families to subpoena Waters to provide more information in court on the submarine mission's findings, including high-definition footage, failed because Lebanon's judiciary was on strike."

"We do not have the footage," Waters stresses. "The footage is owned by [the NGO]. It was actually part of our contract that we don't retain that footage." He shakes his head and stares out over the water.

"The last dive we did," he continues, "was a ceremony where we did a prayer underwater. Uh, it was the first time I, uh, I ever actually buried anybody at sea."

According to Zreika, the Lebanese Army urged Waters and his team not to allow any relatives of the victims on board for the ceremony. "They said, 'We can't guarantee your safety.'"

Waters sucks his teeth, blows out his lower lip. "So, we buried thirty-two people at sea," he says. "And then we put a commemorative plaque on the deck of the boat, and documented all that, and, then, uh, came to the surface," and his voice goes thin and high as he swallows and continues, "and then we did a ceremony on the surface, throwing some roses in the water, or some flowers in the water, and, uh . . . then that was it. We shipped out after that . . . Yeah, so we did this mission."

He shakes his head. "They died holding each other," he mutters. When he stops shaking, he suddenly seems to be more substantial, in focus. He's desperately trying to move on. "The next mission we'll be doing is a volcano mission on a subsea volcano in the Canary Islands... studying this subsea active volcano off the island of El Hierro."

No one exactly knows the origins of the island's name. Many mistake it to mean "The Iron," in Spanish, but it actually derives from the pre-Hispanic Guanche language, and may mean "cistern" or "hero," or a little of both, hinting at the ways in which—in spite of so many wrecks and so many drowned—the water can keep us alive, or, for some, serve as a passageway to a better place, and for those like Waters, give us a reason to live, burst the confines of all the world's Kansases.

Surprisingly, Waters brings up Carl Boyer, his first volunteer assistant all those years ago, back when he built *Trustworthy*. Until a few months ago, the two old friends were still working together in the Canaries, when Boyer decided suddenly to return to Kansas. I can sense the disappointment in Waters's voice, his loss of a touchstone, and a loneliness creeping into that void. "But we're on good terms and everything. He had . . . oh, we had worked together for over ten years." Of his remaining crew, Waters says, "Yeah, we're all kinda scattered . . ." Many of Waters's crew have been leaving lately. I don't want to ask him if it had to do with the taking on of this new sort of work, as they performed in Lebanon. It's so fresh that I don't want to invoke the word *traumatic,* don't want to ask him if he regrets the mission, if he wishes he stuck to devoting his sub to scientific and artistic projects. This month—September 2022—in the wake of the Lebanon mission, he lost his new scientific director of operations too. She quit, in part, to devote more time to what she's now calling her Unnamed Underwater Habitat Project, and—yes—that person is Shanee Stopnitzky.

Stopnitzky had been doing most of her work for *Pisces VI* remotely, traveling to the Canaries only for events like the *Pisces VI* collaboration with the local "hostess" school, Escuela de Azafatas de Canarias, which trains cabin crew for work on passenger ships and planes. Three members of the *Pisces VI* team were featured in a photo that the school posted at the end of the required seamanship training course that Waters helped facilitate: they are Waters (of course), Boyer, and Stopnitzky, uniformly clad in *Pisces VI* navy-blue T-shirts. Behind them, in its spot beneath the marina's canopy, the sub's golden conning tower glows like some downed sun. This is a small, niche world.

This is sometimes a very sad one. Sometimes, a sub bumps up against tragedy, and that sub's captain loses much of his crew as a result. I don't want to ask Waters about this. It looks as if he's already hurting, but of course, that's just my assumption. Still.

Instead, I ask Waters how he made the leap from Kansas to Tenerife. He shrugs and says nothing. Then he shrugs and says nothing again. He clears his throat. He coughs. He's deciding whether or not he wants to tell me something. Finally, he says, "The thing that changed everything was . . . I had been married from a very young age, and, unexpectedly, my . . . my wife had left. And it was a very, very difficult time for me, and kinda everything just changed in my life at that point, where I kinda envisioned my life in one way, then that happened, and I kinda said, well, nothing matters, so I'm gonna take the biggest risks, and do whatever I . . . just whatever. I was maybe pretty careless . . . I decided I was gonna go full force on the submarine."

Waters sold his stock in the family hardware company and put it all into *Pisces*. Indulging his curiosity about living in another country, he found the Canary Islands when researching places that had very deep water relatively close to shore. "I never envisioned it as a thing I could actually see as happening," he says, "[but] I sold my house and I left Kansas."

He widens his eyes. He still can't believe it. "I think there's a lot of times when I realize what I'm doing," he says, "and what, what's happening around me—it's just, it doesn't even feel real a lot of times. It's not . . . not, you know . . . it's just, yeah, not how I thought my life was gonna be. It has been eye-opening . . . And, um . . . it's terrifying, but you definitely feel alive."

Waters runs his hands over his knees, and I realize, though many people have been passing by along the street during our conversation, he has said hello to no one, and no one has said hello to him. Surely people would know him, right? The eccentric submarine guy from middle America? The laurel trees drop their leaves, and one lands on Waters's knee, and he leaves it there. His Spanish, as he puts it, "isn't excellent." From a seaside bar down the street, music faintly snakes— the sounds of castanets and pito herreño whistles, hand drums and what sounds like an orchestra of timples—a five-string ukulele of sorts. At least three voices commingle on a melancholy song called "Isas de Tenerife."

"It was kinda a tragic event that launched me into it but, um, yeah, here I am," Waters says. "It's been, uh, interesting and, and . . . really tough."

And here, finally, the weather drops onto the scene, and a warm wind stirs Waters's hair. I feel as if we're both more present, grounded, and I wonder if Waters, given his predilections, preferred the sense of floating. I try to picture what Waters will never expunge from his head—the bodies of the drowned, still down there, glimpsed, filmed, but unrecovered, ready to become ash, hair lifting like seaweed, still holding each other. The folk band picks up their pace, and one singer's voice rises over the others into this wind, wailing. It is beautiful and nightmarish. Homesick and defiant. Waters and I are doing something we didn't realize we were doing. We are tapping our feet. Down the street, someone is dancing. Someone breaks a string. Waters watches his feet move as if something foreign, a couple of fish. He smooths his eyebrows. A seagull flies into a shuttered restaurant's window, and the song ends, but no one applauds.

"I miss my mom and dad," he says.

# 24

ON THE AFTERNOON of Saturday, August 12, 2017, as the recently evicted journalists stepped outside the courtroom, and as Peter Madsen sat inside, considering the fresh silence after having made out with his own fingertips, a salvage crew gathered at Køge Bay and, at the behest of the police, worked to raise *Nautilus* from the deep. In spite of Madsen's claim that he buried Kim at sea, the police still hoped to find her body on board the sub. The salvage crew prepared for this possibility. They took deep breaths as they stepped into their clammy blue wet suits, deep breaths as they pulled them on, as the material tugged at their leg hairs, as they zipped them up over their chests. The divers sighed as they strapped water-safe flashlights to their heads. Earlier, the crew had been busy on another job fostering the subsea cable connection at Kriegers Flak for the Energinet corporation when their crew chief, Tonni Bering Korreborg Andersen, got the call to abandon that job for an emergency salvage at the behest of the Copenhagen police.

At first, due to high winds and the turbulent chop, Andersen wondered if they could pull off the job, if they even *should* go out to sea or wait for calmer weather. "We have rough seas in Denmark," Andersen says, "so I think the most exciting stuff is to find a submarine because they are so intact compared to other wrecks. With the rough seas, many wrecks are just going to break into pieces over a short time, but a submarine seems to be able to better stand up.

"But even some of the people on board [for the *Nautilus* salvage] from the Danish Navy said this is not possible, you can not salvage a submarine like that [in this weather]." But the urgency of the salvage and the demands of the police compelled him.

Andersen is resolute. He's a big man with expressive eyes, and when he's smiling, the smile is unbridled, and few people appear happier. But when he's burdened, mulling something over, his expression can be almost unbearably funereal. When he says *Yes,* it seems as if anything is possible, expansive as the sea itself. When he says *No,* the world seems to slam shut, and echo. When he's not clad in a thick diving suit, he's usually wearing a baby-blue hard hat and a rugged zip-up jacket. He loves nature, and when he's not diving, he prefers being out in the forest for hours by himself. He loves deer hunting but is quick to tell me that he never shoots more than he can eat—one or two per year. He lives in a thatched wooden house, isolated in the middle of a heath forest in central Jutland, some three hours by car from Copenhagen. He doesn't love the city. It makes him nervous—too much noise and too many people—and when in Copenhagen, he's desperate to return to the woods or the ocean, the peace he sometimes finds there. His hands are enormous, and they sometimes shake. His favorite food is bread, and sometimes, when he's holding a slice, the bread shakes. In 2017, he was fifty-four years old and had already been diving for nearly forty years. Of and in the sea, he had seen so much.

He and his crew of five divers and six onboard salvagers charged with raising *Nautilus* said little to one another, newly aware that the mission was a solemn one. "Peter Madsen, who was in jail at that time, had just changed his explanation," Andersen says. "He said when the submarine sunk, it was empty, right? I don't think he expected us to pick it up. Now he says, 'She might be in the wreck.' Just before [the divers went into the water], we got the phone call from the police onshore that says, something has been changed."

The police milled about far away, onshore, dwarfed by the harbor machines—a massive blue salvage vessel, a green truck with an extended flatbed, and a tall, angular floating crane resembling a robotic praying mantis. Once the divers had pinpointed the location of *Nautilus,* the machines would be deployed to help exhume it.

The bodies of the divers, skinny ribbons of light pouring from their foreheads, grew dimmer as they sank and appeared to one another as murky shadows. To the distant fish, the divers were just another collection of sea crumbs, specks of flotsam in the encompassing brine, waving their limbs, the light from their headlamps wiggly and broken in the tide. Though the divers located *Nautilus* quickly, wedged into the seabed at a depth of only eight meters, it took until late that Saturday night for the crew to raise it and bring it to shore.

"We were now not allowed to empty it of the water," Andersen says, "due to, if there's body parts and things like that. That changed the duration quite a lot. The police told me: every time we have a suspect, if they change their explanation, that turns [the salvage process] a little bit back because then there's something wrong, right?"

Andersen never expected that his attraction to the sea would bring him to this work. As a kid, he was obsessed with Jacques Cousteau. "We had only one channel in Denmark at the time," he says, "and I watched him over and over on television. I think everyone thought Jacques Cousteau was very exciting." Based on his TV preferences, his parents signed him up for a scuba diving class with a local club. He fell in love with it immediately and made excuses to dive as often as possible, feeling more comfortable in the deep than on the surface. "Diving is more or less the best thing that's ever happened to me," he says. Still today, he affectionately calls it "Scuba-Duba."

"Definitely, in the beginning, it was almost like an addiction. I almost need to dive. When you get older, you get more relaxed about it," he says. "Today, I do a lot of exciting jobs and, uh, definitely also a lot of boring jobs . . . Things are not always as you expect. Things are changing, the sea is changing. [I like] the old wrecks, you know, because you can find the pieces down there and figure out the story of what had happened on board."

Andersen confesses that, even after forty years of diving, he still lives in fear of running out of air at depth or that one of his divers will suffer this fate. He feels this fear rise up in him with each mission. "It's terrible," he says. "It happened to me once and . . . that's not a good feeling. That's a thing you don't want to happen to anyone. If you're hours [at depth], like, I was in thirty-eight meters of water, and your scuba [air tanks] are empty. Before your brain realizes they're empty, you need to get up. So it was a long, long way up [for me], and I fainted before I reached the surface, and I was quite sure I was never gonna wake up again. Luckily, I did so. That was probably the most scary thing I've ever tried."

On board the salvage ship, the wind wailing, *Nautilus* still underwater, Andersen and his crew were exhausted. None of them had slept in over twenty-four hours. And before they completed the mission, they would have been awake for over thirty-six hours. The ship leaned its big body—all 2,065 tons of it—against the waves. The ship groaned and the ship creaked, cracked its knuckles and stretched its bones. It was the largest salvage ship that many on the crew had worked on. In a pinch—if a job required multiple days at sea—it could accommodate up to twenty-four in single cabins. The vessel had a gym and a game room and state-of-the-art anchors with a name that sounds like that of a midwestern pinball club—the Delta Flippers. Many of the ship's functions were controlled by joysticks reminiscent of the first-wave Atari system, and blinking square buttons that looked to be cribbed from an old telephone switchboard. It had a main deck and an accommodation deck, a bridge deck and a shelter deck. Some of the decks were painted green and some the faded brick red of a high school running track. Narrow yellow pathways snaked along the main deck, embedded with black arrows labeled CHAMBER, leading into the open doors of blue shipping containers.

The divers and the onboard members of the crew grew so slack from hunger that Andersen had to radio a team back onshore to scrounge up some sandwiches and coffee and bring them out to the salvage boat by dinghy. They ate in shifts, chewed the soggy white bread, sipped the lukewarm coffee from Styrofoam cups, before getting back into the water, before possibly encountering Kim Wall's body.

Members of the crew howled the word *Vina* into the wind, and the wind volleyed it back to them, returning it to their mouths. Out at sea, searching for something, anything dreadful, the divers' hearts were anxious, and this word, *Vina,* this name, too, carried an air of ominousness and mystery, its equitable consonant-vowel-consonant-vowel cluster devolving, via repetition, into meaninglessness, the sort of invisibility that the sea affords the things that have sunk—whether willingly or against their will—to its bed. In this wind, too, the consonants were swallowed, and the vowels were allowed to dominate, screeching *eeeeee-aaaaaa, eeeeeee-aaaaaa* like some ghastly violin or the seesaw of the Tartarus abyss.

The vina is a stringed instrument from India with a lute-like body and a resonator gourd, often used in dirges. And the vina is "Tinder for (girl)friends!" where users can "take quizzes and read awesome articles about living your best life." And Vina was an activist and feminist in Delhi, and the Ultimate Performer of the Philippines, and an attacking midfielder football player from Curitiba. And if only in the *Star Trek* universe, Vina is the sole Earth-born survivor of a doomed expedition to the Talos star group, and after she was "repaired" by the Talosians, she was groomed to be an Eve figure, once the Talosians found a suitable "Adam," and they held her captive until she began to suffer from Stockholm syndrome and, so suffering, began her most memorable monologue: "I'd been alone for so many years, I never imagined happiness or love . . ."

And during antiquity, Vina was a city and diocese of Roman Africa that still houses the ruins of an amphitheater dedicated to Marcus Aurelius, the final emperor of the Pax Romana and, therefore, the gateway from relative peace to bloodshed. And Vina is the name of the savior goddess who never was.

But this is not antiquity, and this is not something that never was; this is here, on board M/V *Vina*—the red-and-white salvage ship that Tonni Andersen is captaining in the Øresund Strait. And on M/V *Vina*—this *Vina*—some of the crew intone, "Vina!" as if riders trying to steady a horse charged with crossing a dangerous ravine.

And some aboard *Vina* shout, "*Nautilus*!" and make of *Nautilus* a sea monster again, distant cousin to the giant squid—the "devil-fish"—shooting through the water by means of jet propulsion. Down there, at depth, with the water moving just so, nothing is fixed, all is dynamic and animated—the beautiful is terrible, the terrible, beautiful—and the divers and salvage crew, so intimate with the deep, know it deeply, this notion of simultaneity having colonized their brains and bones. The beautiful terrible. And beautifully, and terribly, the leviathan rises from the deep to meet the shouting of its name. In a plume of light, sheen of steel, and boil of water, the DIY submarine ascends its chamber of water. The divers swim frantically away toward a less agitated area of the deep.

A blister forms on the water, haloed in foam, turquoise and sulfurous like some congested geyser desperate to dislodge some toxin. The onboard crew spread their feet, bend their knees, and take quick shallow breaths, readying their bodies for the pop. *Nautilus* forces itself against the surface of the strait. It breaks the skin. *Vina* rocks nearly to capsizing. The crew's shouts are indistinct now in the chaos of spray and sea-roar, but they stay on their feet. They stay on their feet and do not fall as they lasso their straps around the submarine, bobbing now, both gracelessly and full of lost grace. Defunct and enormous, impressive and pathetic, it appears as some ancient bomb still capable of arousing its detonator, an animal, dead or playing dead, stilled, but prepared to strike if so provoked, at least in the imaginations of the crew who hold their palms to its cold metal flank, and to the divers who now rise from the depths into the surprise of night, squinting against the hard gleam of the moon, reflecting from its conning tower, a mischievous lighthouse, blinding rather than warning, tempting all who come close to wreck their bodies against the rocks.

Andersen's heart was racing, and, somewhere in his body, he still harbored the fear of running out of air.

"We got the submarine to the surface," Andersen says. "But," in spite of the sturdiness of the thirty-four-ton pedestal crane anchored to *Vina*'s aft, the vessel's seventy-five-ton heavy lift boom, and six badass warp winches, "we were not allowed to lift it on board; not able to lift it on board. So we decided to tie it on and take it to the nearest harbor with big crane facilities . . . We had about eight hours at very slow speed to harbor." And during those hours, some of the crew tried not to think of what the sub they were towing might contain. Some were so exhausted, they forgot that they were towing a sub at all and forgot where they were, bobbing in some liminal watery space between asleep and awake. A couple of them actually dozed on their feet as the stars winked and fell and set each other on fire.

Off *Vina*'s stern, *Nautilus* bobbed like a bath toy. But at the harbor, the sub emerged in its entirety, dangling by only two reinforced orange straps from the bent neck of the harbor crane, caked in silt, the water dripping from it in strings, and seagrass bearding the blade at the bottom of its outer hull. The sub appeared as the bloated body of some ancient and indiscreet mer-god, defrocked and condemned for some bygone abuse of power. The diving team emerged too, stepping from *Vina* to the dock, looking much like *Nautilus* itself, only limbed and in miniature. Their hair and their beards hung slack and scummy, and their faces and necks were bright red. Some couldn't stop rubbing the water from their eyes. Those who still had the energy to shout instructions to their mates did so in hoarse, water-clogged voices, their veins jumping like cello strings. They seemed as if animated by some phantom mad composer, their bodies being played and bowed and struck by some invisible puppetmaster. Everything about them seemed atonal, another avant-garde orchestra playing another tough-to-take symphony amid the weird industrial art scene of Refshaleøen, just as the scene, and the island itself, was busy losing its innocence.

When the recovery of *Nautilus* was complete, and the police took over, the exhausted crew of *Vina* had to return to the job they had abandoned at the urging of the police—what Andersen might call the more "boring" labor of laying subsea cable. And though the job was done, it wasn't done with Andersen.

Tonni Andersen is still wrestling with what he now knows about the case as compared to what he knew then, as *Vina* chugged through the chop and cold wind back toward Copenhagen. "At the time, we didn't know what was happening," he says. "Nobody knew that [Madsen] was, you know, totally freaked out. And we just thought [Kim Wall] *might* be in the submarine, but the story then was she fell down some of the stairs and hit her head. The way we got [the story], we thought it could be an accident. We didn't know. But afterwards, yeah, it changed me. I mean, how crazy can you get?" When pressed as to how the experience affected him in the aftermath, Andersen gets, strangely for him, noncommittal. He takes a long pause, then says, "uuuuuuuuhhhh, mmmmmmm," then pauses again, then says, "aaaaaahhhhhh," as if enduring some dull pain, then shaking it away as if cobwebs.

"The guy [Madsen] was, in any way, brilliant," he says when he finds his words. "He invented submarines. He invented a rocket. I think he was brilliant, but in the brain, he was totally fucked up, right, because then he did something like that, so that shows there's a thin line between being brilliant and then, you know, going totally mad. Ugh. Ugh.

"But the sea?" he says, turning away from the ugliness. "I appreciate [the sea] more and more." And it seems as if he's working something out, trying to reconcile his love of the ocean with all of the unlovable things he's had to encounter within it. "I mean," he says, "nobody throws garbage into the sea anymore. I mean, the sea is really important for all of us." He says he has no plans to stop working, says he wants to work into his seventies, maybe beyond. "When I think about stopping," he says, "I think, *Oh, no.* [On board a vessel] you have so many beautiful sunsets at sea when the weather's good, and you can't see them anywhere else. When you stop sailing, you're not going to see them again."

And then, without prompting, he says, "But the worst is when you have to salvage bodies from ships that have sunk, you know. That's not a good experience." He says he has unfortunately participated in numerous such missions, and, for a while, he was able to let them slide off his shoulders, but, for some reason, after the raising of *Nautilus,* and all he found out about the case in the aftermath, the horror of those former missions came rushing back. Since, he's had trouble sleeping.

When he wakes in the middle of the night, he hauls himself out of bed and paces the rooms of his house in the middle of the heath forest. To try to calm down, he'll go to the living room and listen to one of his favorite records from his collection of over seven hundred LPs—Led Zeppelin or AC/DC or Johnny Cash or Elvis Presley. Sometimes, he'll pop in the tape of his favorite movie, *The Shawshank Redemption,* and watch until he unwinds. And sometimes, he'll go to his Lego room and stare at his collection—towers of boxes rising from the floor to the ceiling; there's hardly room left for him in there. For years, Andersen has collected Lego's Monuments of the World series, but he's never built them; never opened the boxes. He's saving them for retirement, so he has something to fill his last days—building little models of the Eiffel Tower, the Colosseum, *Titanic.* He estimates he has so many that the Lego projects will occupy him for over ten years. But for now, sleepless in the middle of the night, he'll stand wedged in among the boxed monuments-in-pieces and imagine one day building them, and he'll breathe until his heart slows.

"It gets to you," he says, "definitely. Once, I salvaged five [bodies] from the same boat—a [Scottish] fishing boat in Danish waters, and it went down in about thirty-five meters of water, in extremely bad visibility, and I had to go into the wreck and into the cabin and down to the bottom and try to find these people and . . . you see their faces. Now, years and years afterwards, I've started dreaming of those guys in the night . . . I could see their faces coming up. It's difficult times . . . They're not unfriendly, not scary, just, ah, uncomfortable, you know? You see their faces, and they come back to you . . . I have to accept that they're there. They've started coming back to me in the nighttime, and I say, *Hello*."

# 25

So: I can't swim and I'm afraid of the ocean, and I'm about to dive to two thousand feet in a home-built amateur submersible off the coast of Roatán, Honduras, in the hopes of spotting a giant Triassic-era sixgill shark feed from a slurry of fish and goat viscera that the amateur submersible builder and captain Karl Stanley pre-dropped into the sea for my benefit the night before. Sometimes he weights and tosses a pig carcass down there; sometimes a deer or a horse. "It was pretty rank," he says, "fermenting in its own juices on the dock. The more the flies like it, the more the sharks like it. Even after I popped it in the chest freezer—a lot of flies."

His neighbor, "a notorious narcotrafficker," Stanley says, is blasting "Because I Got High" on repeat. Stanley has to shout to be heard. He struts to his yellow submarine, scratches it where its head should be. "The vehicle I'm operating," he says of *Idabel,* the claustrophobic, nine-thousand-pound, three-person steel can I'm about to board, "is the deepest-diving manned vehicle in the Western Hemisphere south of the U.S." He bounces on the balls of his bare feet as he says this, rocking on his long toes. His fingers are long too, and tan, and the nails are bitten to the point at which they appear painful.

His dock is littered with wayward wires and bolts and straps and blades, beneath a canopy of stretched-out sails. The water is shamelessly turquoise, the clouds feathery; the fat palm fronds clack like castanets. When Stanley picks up his phone, he answers it, "What up?" His dogs, Doris and Mishka, race around the dock, nearly knock us over. Stanley yells at them. Last night, I met Stanley for dinner at Loretta's Island Cooking—a seafood and chicken shack—and he had his dogs in tow.

Last night, there were three of them—Doris, Mishka, and Kujo (with a K, like Karl), the latter a sweet old pit bull whom Stanley claimed as his favorite. I had arrived at Loretta's first, and through WhatsApp, Stanley sent a message suggesting the garlic conch dinner and asking me to order him the same; he'd be there in five minutes, he said. When he arrived a half hour later, my conch was done and his was cold. He sat down and took his first forkful before saying hello. He spent most of the dinner yelling at his dogs.

Now the sea breeze stirs our hair. Stanley cracks his knuckles. Mishka tackles Doris and Stanley claps his hands and screams, "Hey!"

"Where's Kujo?" I ask.

"Kujo's dead," he says. "Hey!" he screams again at the dogs.

"Wait, what?" I say.

"Kujo's dead," he says.

"How?"

"She crawled off under the neighbor's house last night and died."

Stanley, sentimentally, tells me he dragged Kujo's body out and carried her to his wheelbarrow and cut the dog's head off with his hacksaw. The fireworms, he assures me, will eat it to the bone. Then he will mount Kujo's skull next to the horse pelvis on the front facade of his house.

"There was an open space there, so . . ." He shrugs. "And those worms are crazy. Should be ready tomorrow. Lots of things like that around. I was followed here by one of the island's infamous snakes."

The crust of last night's sleep holds to his lashes. Stanley projects a cocktail of sentiments at once—anxiety and calm, confidence and agitation. He's enduring me, and he's equipped to do so until the end of time. When he speaks, he does so through clenched teeth, and his voice is high-pitched and seems prerecorded. There's an echo to it. He sounds like the smartest child in the room, intriguing but also menacing—Big Bird on MDMA. I focus on Stanley. I try not to look around the yard. I don't want to see that wheelbarrow.

"A corn snake," Stanley continues, "big one. We've crossed paths before. I recognize him because, like, three inches of the end of his tail is missing. Something got him." His voice trails off. "Maybe a machete," he mutters. He pats *Idabel*. He's been living on Roatán for nearly twenty-five years. He knows a lot of snakes. He is forty-eight years old, and he's six feet tall, but his hair, billowing into the wind, makes him seem taller. His hair appears as if the port of some tractor beam is about to descend from the heavens, whisk him into the clouds.

I consider my lungs. I stare at *Idabel,* the submarine so little it can fit in the back of a pickup truck, and take inventory of my ailments. I'm worried about my asthma, my high blood pressure, my anxiety attacks; I'm worried the retina in my left eye will detach with the pressure and I will go blind like my late reviled grandmother. I'm worried about my small bladder and my frequent urination because *Idabel* harbors no toilet. I live in a near-constant state of pee anxiety, and I've had a lot of coffee. This dive is going to be three hours. I imagine my bladder bursting. I imagine the barracuda threading the plume of discharge, bemused, before the ocean sweeps it away.

I worry that the hands into which I am about to place my life may not be interested in carrying it. To Stanley, I'm just another obstruction, blocking his view out the porthole. He's been diving in home-made subs for twenty-six years. Shanee Stopnitzky, who's familiar with Stanley on the personal submersible circuit, told me, "Karl is a fucking weirdo. He's amazing. He's, like, definitely a character . . . He built that sub himself. Oh shit, oh yeah. He's, like, *the* person."

She told me that the PSUBS community worships Stanley, though Stanley never shows up at any of their conventions or contributes anything to their online chats. But they talk about him as if he's a celebrity-ghost, some hermetic, Salinger-like legend haunting their niche world. They create chat threads on their mail list with titles like "Remember Karl Stanley?" as if they're not quite sure if Stanley is actually real and still living. They memorialize him while he's still here, shifting his phlegm, picking his toes, tinkering on his submersible. Some speculate that he briefly contributed to their chats, under the pseudonym of Captain Nemo, but others disagree, and heated debates ensue over Nemo's true identity. Nemo's few posts extoll the virtues of his sub, *Nautilus,* which some insist is a pseudonym for *Idabel,* and causes others to speculate on Nemo's relationship to Peter Madsen. "[Nemo and

Karl Stanley] *must* be the same person," one contributor desperately posits in a post that overuses the word *marvel*. "Let's just call it a mystery, alright?" suggests another.

Compared to Stanley, Stopnitzky said, most of the PSUBS folks focus their attention on the mechanical side of the subs, and they don't end up diving very often. Instead, they geek out over the gear, and fetishize the parts, and talk about what will work and what won't work at depth only in theory. "They, like, know everything about the systems, and that's what they're passionate about," Stopnitzky said, "but they never fucking dive! But Karl: he was like one of the first people in that community to build his own sub, and he's done more dives than just about anyone else, like, in any submarine."

According to Stanley, there are about one hundred home-built subs in the world, and only a tiny fraction of them are currently diving. Very few of those active subs dive past a hundred feet, and very few do more than one dive a year. Stanley, meanwhile, dives to two thousand feet roughly a hundred times a year.

"By most metrics, he's the most successful submarine person that there is," Stopnitzky said. "He has a very serious love affair with the deep. He has this crazy operation that he runs out of his personal house. It's incredible. The whole thing is incredible."

⁓

Last night at dinner, the first time we met, not ten minutes after Stanley showed up with his three dogs, he asked me, "So, do you know about that Danish guy who murdered the journalist in his submarine?" Both Stanley and I were acutely aware that I am a writer and that I was planning to dive with him on his sub the next day. My throat went dry, and I sipped from my water. Stanley stared at me and drank from the opaque thermos he brought with him, filled with what he called "medicinal lemonade." We sat across from one another on the outdoor deck. Something screeched in an adjacent palm tree, and a coconut fell. Kujo—still alive and whole—nosed my palm under the table. I told Stanley, yes, I do know about the murder.

"Oh yes?" Stanley said. At a neighboring table, silverware clanged too loudly against a plate. "He knew exactly how it was gonna play out.

I mean, I never met the guy, but everybody [in the personal submersible community] was aware of it when it happened. And we got a lot of press. But pretty much every profession has somebody that's been a murderer, right? No one was really surprised. He'd been well known on the S&M scene, talking about torture and murder abstractly for years. So we weren't like, *Oh my God! Not him!* He did it for the thrill. To see if he could do it. Once he had her on the sub, she was trapped." Stanley took a forkful of garlic conch and nodded at me, wide-eyed. "Yeah," he assured me through the mouthful of food.

"So, you're not into any of that, are you, Karl?" I dared.

A bell rang in the kitchen. Somebody's fried chicken was ready.

"You're totally safe," Stanley said. "A journalist in a homemade sub? A journalist in a homemade sub murdered? I could never murder you under there because it'd be, like, everybody would assume I was trying to be a copycat. Everybody would think I don't have an original thought. That would be the worst part. I'm not a copycat."

The remainder of the meal was pocked with great silences. The three dogs played; the kitchen staff laughed with one another. The ocean roared. We finished eating, and Stanley told me that he'd see me tomorrow.

That night, I did not sleep, and because I did not sleep, the only nightmares I had were the real ones.

~

Because I got no sleep, I walked the streets of Roatán to Stanley's place earlier this afternoon as if through a fresh dream, passing the Rosita butcher shop's bloody windows and the acrid smell of powdered soap snaking from Cindy's Laundry. Seabirds screamed and fought along the courtyard stairwells of the Arco Iris motel. Someone urinated in the alley next to Woody's Grocery, where six cats prowled the stack of pineapples at the outside display. Men peddled brightly colored burner cell phones. Women peddled massages. Berinche Street dead-ends at West End Road, which runs along the ocean. The broken tawny glass of countless Salva Vida beer bottles decorates the roadside dirt, catches the sun. The foliage is green and unruly and lurid. A gecko called to me in a hoarse voice. *Bull*-shit, *Bull*-shit, it said. I hung a right and found Stanley's place past a charter boat service called Ruthless Roatán. I was told to look for a house with a steer's skull and a horse pelvis mounted on the facade. What I wasn't told is that the steer's skull would have pink Ping-Pong balls in its eye sockets, lit from the inside so they glowed.

Stanley was waiting on the patio, flanked by a tangle of palm fronds and seagrass. His house is in part made of a dead coral reef—the entire ocean-facing facade of it, the crooked pillars—some condemned castle of bone. He's recently begun renting out rooms on the first and second stories, and so his living quarters consist of a cavern that he cut into the rock beneath the house itself, which means he lives underground, one foot above sea level. Though he's been living in this house for the past eighteen years, his quarters remain predominantly empty.

He sat on a cushionless white stool with short legs. His bent knees rose to his ears. There was something ascetic about his posture, and I'd associate him with a monk if he weren't wearing baggy shorts and spreading his legs. Though we had already many times discussed his fixation on the deep as an "obsession," today Stanley decided that the term didn't apply in his case. He whipped out his cell phone and showed me the Vocabulary.com dictionary entry for *obsession* as his evidence. "I don't think 'obsession' is the right word," he told me. "I'd say that if you want to be successful at certain tasks, you have to be 'highly focused' or things won't turn out well."

The breast pocket of his gray T-shirt was played out and saggy, as if he's stored all manner of nuts or bolts or the seashells he collects inside of it. He scratched at the stubble beneath his chin.

"By the way," he said, "I was thinking more about the whole Peter Madsen thing. You're totally safe. I'm not a fucking cliché."

My heart raced. I was scared; worried that Stanley was more concerned with being associated with *cliché* than with *murderer*. And I'm not sure what my face did here, but it must have done something, because Stanley felt compelled to say, "C'mon, c'mon. When else would I—a submarine builder—get to talk about this with a journalist who's about to go down in my sub?"

≈

Stanley's two remaining dogs chase each other into the foliage and do not reemerge. Stanley preps *Idabel,* cleans the portholes, the largest of which he found in an old warehouse in the Bronx, sealed untouched into a crate like in the last scene of *Raiders of the Lost Ark.* "This window's older than me," Stanley says. He squats and rises, and his knees crack. He tosses the red rag toward a blue plastic barrel at the dock's corner. He misses and commences bobbing up and back on his toes, ready to pounce, it seems, but slowly.

As yet, I can only imagine Roatán's underwater world, which is reputedly inhabited by the vengeful mythological sea monk who curses those ships that arrive with colonial ideals, however latent; seeking retribution against the descendants of Christopher Columbus, who landed on the Bay Islands (of which Roatán is a part) during his fourth voyage in 1502, bringing with him epidemics of smallpox and measles. The Spanish heaved the dead into the sea, where they were eaten by the sharks. The few members of the indigenous population who survived the diseases were enslaved by the Spanish, who decimated the original native communities here.

In 1638, the British arrived and occupied Roatán's east end, igniting a two-hundred-year conflict between the Spanish conquistadores and the British pirates, both of whom wanted control over the island's resources and its position as a way station for their trade ships carrying silver and gold raided from the so-called New World back to Europe. They engaged in proto–trench warfare in the mangrove tunnels. In ensuing years, the French and Dutch also established settlements and refreshment stations on the island. In 1797, the French battled the British for control of the isle of Saint Vincent, which was home to the Garifuna, a people of mixed free African and indigenous American descent. The French enlisted the Garifuna to fight the British on their behalf, and when the British defeated them, they deported the survivors to Roatán, where many remained, becoming the island's first permanent settlers since the extermination of the original indigenous inhabitants. After the passage of the 1833 Slavery Abolition Act, many of the formerly enslaved left the Cayman Islands for Roatán, seeking a fresh start, away from the trauma they had endured across generations. Many former enslavers from the Cayman Islands also settled along the pristine seaside on Roatán's west end, where Karl Stanley's house and sub sit today.

Some believe that the fabled sea monk, deep-sea-dwelling protector of these waters, is still hatching further retribution for the sins of these colonizers, enslavers, and their descendants, those still disguising their atrocious missions in terms like *pioneering, stewarding, settling;* the manifesting of their "manifest destiny" depending on the snuffing out of all other destinies. Still today, the island's eastern half—where the oldest Garifuna communities are—is being ransacked by overseas corporate interests and their campaigns of overdevelopment.

As news of the sea monk spread with "viral-like efficiency," some of the early British and Spanish occupiers likely claimed to have seen such a creature from the decks of their ships and as a result believed Roatán to be protected by the devil, a damned place. The sea monk was rumored to be relegated to the waters of Europe, and so it was seen here in the Caribbean as another anomaly, abomination. It wasn't supposed to be here. Today, some naturalists postulate that the sea monk was actually a monk seal, the Caribbean species of which, now extinct, was once—in the time of Columbus, and in the time of the British and Spanish occupation of Roatán—abundant in the waters off Honduras. (Columbus had in fact also claimed to have seen mermaids in the Caribbean, now believed to have been manatees.) Still today, according to the Convention on Migratory Species, every once in a while, an endangered Mediterranean monk seal will migrate unpredictably and turn up off the Honduran coast, and for a moment one may think that the extinct has risen from the dead and that vengeful legends may indeed be real.

Some believe it was the sea monk that caused the offshore wrecks of such ships as the *Aguila* and the *Odyssey,* in whose broken holds and staterooms silver fish now sleep and hunt and mate and make plans. Some believe that the sea monk may be responsible for the Lluvia de Peces miracle—a yearly occurrence wherein a deluge of fish rains down from the sky onto the mainland town of Yoro, Honduras. This phenomenon has been occurring for centuries, and scientists have tried to explain it via meteorological reasons, citing the intense winds and waterspouts that can sweep thousands of fish from the ocean and carry them into the sky. But many locals argue with this, wondering how exactly the fish are carried all the way from the Atlantic, forty-five miles inland, and dropped—every year—precisely onto Yoro, where the Spanish priest Father José Manuel de Jesús Subirana, one of the more aggressive spreaders of Christianity here, based himself in 1855 and launched his campaign of conversion. Today, some who view Father José favorably believe that the Lluvia de Peces is a reminder of how the priest took pity on the "poor" and, using his power of prayer, compelled the fish to rain down upon them, providing them with food. Those who take issue with the priest's colonial agenda see the rain of fish as a mockery of him and his actual inability to manifest miracles. Instead, the Lluvia is a gift from the spirits of their pre-Christian ancestors—the indigenous who were so long ago wiped out by the occupiers and slave raiders.

I watch the sky for fish—for miracle or mockery, plague or reassurance. Any magic—natural or conjured—to distract me from what I'm about to do, show me that the world is so much bigger than the cramped yellow sub into which I'm about to lower myself. Stanley brags about the virtues of being a self-taught engineer. In doing so, he compares himself to Christopher Columbus. "I don't think you have to have an education in anything to be an explorer," he says. "You just have to be curious enough . . . What did Christopher Columbus have a degree in?"

*Idabel* bobs aloof in the water, and Stanley sways on his feet like a tree. Somewhere, beyond the dock, the carcass of a headless dog clots in Stanley's wheelbarrow. I don't ask him about the sea monk or the yearly downpour of fish. I'm afraid of what he'll tell me or afraid he'll laugh at me. Instead, I stare into the water, and through the sun's glare, everything within remains invisible, occupying, however in my head, both the realms of the real and the mythological. Until we drop through the water's surface, everything down there can remain in the dreamworld. I can calm my heart by telling myself that none of this is entirely real—not yet. For a brief second, I can't quite believe that fish—let alone a torrent of them—actually exist.

I mimic Stanley, rocking on the balls of his feet as if to keep something insidious at bay. Strangely, as I start, he stops. With his long toes, he drags a bathroom scale from beneath a workbench. He weighs me. He tells me to take off my shoes, leave them behind. With a whipping of his arm and a groan that may come from him or from *Idabel*, or from both, Stanley yanks open the hatch. A seagull shits onto my left sock. Its shadow passes over Stanley's face. I'm waiting for him to show some sign of affection or camaraderie. I'm waiting for him, I realize, to hug me. The seagull moves on. Stanley opens his mouth. This is going to be grave. "Climb in," he says.

≈

At age fifteen, Stanley bought his first piece of steel and, fulfilling a vision he always remembers having, fashioned it into a winged submarine in his parents' backyard in Ridgewood, New Jersey. It was a design unlike any other—a hybrid cobbled together of various forms of inspiration, from books on submarines to fever dreams. It was meant to glide through the water without motors. And it did. The contraption worked. With it, he made 556 dives over two years. He spent most of those years underwater. The machine inspired in him a dreaminess that made life amid the "surface institutions" seem frivolous.

He had already been spitting in the face of those institutions whenever possible for most of his childhood. No matter what he did on dry land, though, he couldn't quiet his restlessness. Not that he didn't try. Though *Money* magazine named Ridgewood among its "Best Places to Live," Stanley disagreed. He scoffed at the town's pride and joy—the thirteen properties listed on the National Register of Historic Places, including the historic Graydon Pool, in which Stanley frequently urinated. He hopped freight trains and rode them across the continental United States. He broke into the local police station and destroyed the file containing the parking tickets. He made pilgrimages to the ocean and back-floated in deadly rip currents. Unsupervised, he journeyed to Coney Island to behold a 1930s-era bathysphere, a steel orb that was once lowered on a cable to explore the deep waters off Bermuda. Barefoot—always barefoot—he developed a habit of climbing. He scaled telephone poles and stadium floodlights. He once climbed a radio tower over a thousand feet tall so he could be closer to the point at which the signals escaped, blipped upward toward the cosmos. He wanted to be part of that conversation. He wanted to immerse himself in the invisible existent, in this case the sea of static and shock, and all of the other voices it carried.

"I don't miss the States," he says. "At a young age, I saw my dad put on a suit and commute into New York City and—my dad's told this story a bunch of times in front of me—I would say, 'I'm not gonna be like you, Dad.'" His parents couldn't handle the manifestations of his restlessness, and when Stanley turned fourteen, they packaged him as a delinquent and sent him, like Jacques Cousteau before him, away to reform school.

"They had this team of private detectives manhandle me into a car and drive me off to Maine," Stanley says.

Isolated in the Maine woods, twenty-five miles from the nearest town, lorded over by headmasters, Stanley knew upon arrival, after his shoelaces were confiscated, that a traditional escape plan was out of the question. He hatched an atraditional one. "My strategy became to get on everybody's nerves, so everybody would want me out of there. So nobody would be advocating for me to stay. Knock over furniture if I had the chance. But there was a key thing to let them know I meant business. A shower strike. That's my ace." For weeks, Stanley refused to bathe. He would wake in the night and scream and scream and scream. And so, also like Jacques Cousteau, he succeeded in getting expelled from reform school. His parents, convinced he needed treatment, refused to take him home and instead committed him to a state mental hospital for six weeks where he was diagnosed with what he remembers being "defiance-of-authority" syndrome. "They tried to give me medication," he says, "but I refused. Poured it down the drain."

Stanley was able to rekindle a relationship with his family via a "process" that was accelerated by his mother and sister visiting him in Roatán and going on a sub dive together. "We had a sixgill [shark] come by," Stanley says. "The first time I had a horse as bait was when my mom was visiting. She was there in the morning when it got shot.

And she cried seeing the horse die, but she had a picture of that shark as her screensaver as long as she had that computer. So yeah, she was proud of me."

≈

Upon his release from the mental hospital, Stanley narrowed his focus and conjured the resoluteness, born of the anger of having been committed, that compelled him to think of nothing but submersibles and to learn how to build them. He had been nursing his obsession with subs since age nine, but now, at fifteen, he became serious. Immersing himself in this obsession became another way to isolate himself from the surface world, and the obsession itself became a protective enclosure within which he was in control—a metaphorical sub, before he built the actual one. His obsession hardened into what he might rather call a state of being "highly focused."

"What I love most about being in the sub," he says, "is you're in a place where there's zero probability of running into someone. It's a frontier, and you're totally on your own. No Coast Guard. No air traffic controllers. No one can control you."

Fifteen-year-old Stanley convinced his parents to buy him welding tools, and they were ecstatic that their son was expressing himself in a way that might lead to his landing of a job one day. He saved up his money from working at the local ice cream shop and bought that fateful piece of steel—a ten-foot-long, quarter-inch-thick pipe. When it was delivered to his parents' house, the truck driver was surprised that Stanley was so young and had no tools ready to transfer the pipe from the flatbed to his backyard. But Stanley, who had daydreamed through shop class in school and had previously exhibited little engineering prowess, called up his colleagues from the ice cream shop and designed with them an intricate hammock comprised of a bunch of old dog leashes. With this, he was able to maneuver the pipe into the backyard, where, over the next three years, culling inspiration from submarine books and magazines, museum dioramas, and old documentaries, he planted himself in the shade of an apple tree and, in a torrent of sparks, with his welding torch, shaped the vessel. The birds complained, and the apples fell. The neighbors gave his parents puzzled, concerned stares, but they shrugged these off, happy their son was devoting his energy to something other than climbing into the sky on power grids. Stanley christened his vessel-in-progress *C-Bug* (for Controlled Buoyancy Underwater Glider). His parents called it "the pipe in the backyard."

He spent the next seven years working on the thing, finishing it just as he graduated college. The pipe made up the majority of *C-Bug* and was able to accommodate the legs and torsos of two people, their bodies pressed together as if slow dancing in place. The passengers' heads were helmeted by two small "towers," each affixed with a Plexiglas porthole. The steel wings Stanley welded beneath the ballast tanks allowed him to leave out an engine entirely, as *C-Bug* was intended to glide through the water like a raptor on a thermal. Stanley loved that the lack of engine noise would allow for a greater communion with the sea—finding a peace and a holiness in the silence. He wanted *C-Bug* to be as quiet as just another fish.

He staged the launch of *C-Bug* at a rank little creek on the outskirts of St. Petersburg, Florida (where he had gone to college). About a hundred people gathered in the crabgrass to watch, slapping the mosquitoes at their necks. They expected grandeur, and Stanley did not disappoint. He climbed in and whirled his pointer finger in the air. He closed the hatch and dove all the way to the creek's bottom, a depth of twelve feet. The crowd cheered, but Stanley couldn't hear them. He stared out the porthole at the scum and the mud and the beer cans and the dirty diapers, and the occasional fish that had adapted to live among them. He knew this was just the beginning and imagined the creek bottom fissuring beneath him, swallowing *C-Bug* into the uncharted depths of our world, where our myths and our monsters become actual.

≈

To have a sub professionally tested for safety and design flaws is expensive, upward of $15,000, so, following the maiden launch of *C-Bug*, Stanley, fresh out of college and money, proceeded as many backyard sub builders proceed—via trial and error at depths wherein a leak or a gasket extruding, or an outrigger imploding (all of which *C-Bug* endured), wouldn't necessarily mean certain death. Then—as is his penchant—he took the trials and the errors deeper. After graduating, he stayed in Florida and spent nearly half of the next year underwater. He had *C-Bug* towed farther and farther into the Atlantic, where he would be harassed by the Coast Guard, who needed assurance that it was only his own life he was risking. U.S. policy dictated that if he wanted to risk other people's lives, he would need to cough up $100,000 for the proper license to do it. Still, the Coast Guard struggled to find a rationale for why he wasn't allowed to take his vessel down. Being unmotorized, under twelve feet, and noncommercial, *C-Bug,* Stanley argued, fell into the same regulatory category as a canoe, and therefore needed no license.

But Stanley was already conceiving of ways to monetize his dives. He turned his ears to the wind, to the whispers thereon. At a dive show in the summer of '98, one of these whispers—uttered by a colleague of Stanley's—spoke of a place called the Inn of Last Resort, a sea-battered hotel at the end of a dirt road on a duodenum-shaped peninsula on the isle of Roatán, home to the brilliant fringing reef system rising up from the abyss of the Cayman Trench, estimated to be twenty-five thousand feet deep. In Honduras's waters, Stanley wouldn't be subject to regulation, wouldn't need any kind of paperwork or license or certification. He wouldn't even need to insure his craft.

This, Stanley thought, sounded like his kind of place—the fantasyland he sought years ago when scaling those radio towers. This time, neither Coast Guard officials nor his parents could commit him to an institution for his perceived transgressions. This time, he was determined to forsake the strictures of our surface institutions for good.

The Last Resort is an inn that's shed more of its baby-blue paint to the flora than it's retained on its wood slats. The place is overtaken by plants that look painful to touch—that could cut one to the bone—and produce flowers beautiful and toxic, carnivorous and conspiratorial. In the surrounding jungle, the monkeys binge-eat the hibiscus. Here, Stanley started his hustle, peddling dives on *C-Bug* to depths unattainable by scuba endeavors, working for the Last Resort's owner, a former trainer of the 1980s-era Honduran contra death squads, who was happy to provide his guests with this new and unique excursion experience. Stanley found, at depth, that people treated his sub as a confession booth, admitting their sins and regrets, seeking absolution from the sea or the barefoot captain working his levers, spread-legged and aloof. Speaking of the depths, Stanley would utter cryptic but earnest maxims. "Nobody can reach you," he might say, staring at the deep blue and the orange coral, but seeing perhaps that springy boarding school cot in the Maine woods, on which, itchy and unclean, he once screamed and screamed into the night.

 Stanley loves it when he's at depth and it begins to rain, the drops dappling the surface hundreds of feet above him. To be submerged is to be *under,* but to be submersed is to be both *under* and *within.* When it rains like this, Stanley loves being so far beneath one kind of weather, and immersed within another—the weather of the brine, a weather beholden to a different set of elusive rules. Sometimes, that weather surprises him. And even when that surprise is harrowing, he's thankful for it, thankful for the chance to regroup and adapt to it. On *C-Bug*'s early dives, small disasters struck. The porthole cracked and water sprayed in. The sub's electrical panel caught fire. Each time, Stanley was able to surface, where he and his customers, upon seeing the sky, and trees, and maybe rain, and maybe the sunset, described the bursting through as a rebirth. Some people vowed to change their lives. Stanley vowed to address the issues and dive deeper next time.

 He was fixated on uncovering his sub's boundaries, its line; to test how deep it could go without imploding under the pressure and crushing everyone within it; to test its parameters and his own. To see what it, and what he, was made out of; what it, and he, could take without breaking. He found that the deeper he went, the more beautiful the seashells he found, which he started collecting and selling to tourists on the side. He spent some of his nights on the sub, anchoring it on a coral ledge, hoping the sharks that bumped against it wouldn't knock it off. He got squirrelly. *C-Bug* couldn't take him to the depth to which he dreamed of sinking. He retreated into his head, which was ornamented with visions of an all-glass spherical submersible that would dive to ten thousand feet. Like Shanee Stopnitzky, he envisioned fashioning an underwater house, then an underwater pod-shaped hotel for the wealthy honeymoon circuit. When asking about the safety of testing his sub's limits, it's as if one is asking him about the safety of testing his dreams, and he grows aggravated, crossing his arms and staring off over one's head.

While engineers for the Navy can assess the safety of a "professional" sub, there's no official way to assess the safety of a DIY sub like Stanley's. The DIY sub isn't a monolithic entity, varying in structural soundness among builders. It's situation-specific. When questioned about the safety of an amateur submersible, Navy engineers tend to respond with some variation of, "Listen, there's no way I'd get into that sub." If Stanley pushed his sub too far, or if one of those curious sharks knocked it off the undersea ledge when Stanley was sleeping, and it descended a few inches past its depth capacity, one Navy engineer describes the hypothetical catastrophe as akin to the crushing of a soda can.

In 2002, after spinning his wheels shell collecting, doing salvage work for the Cuban government—exhuming anchors and amphorae, and searching for sunken Spanish galleons rumored to be loaded with gold—and sleeping five hundred feet under the sea, Stanley met an American businessman and machinist who offered to rent him an airport hangar in Oklahoma. He was frustrated by *C-Bug*'s meager operating depth of six hundred feet (though he once, in a depth-crazed state, risked his life by pushing it to 725, which permanently deformed the vessel's hull).

"Why not go farther?" he asked himself. He became obsessed with finding the seafloor in a steep area of the Roatán coast where one of those official Navy ships had sunk a few years prior. Stanley took the machinist up on his offer to break the stagnation. He made the temporary move to Idabel, Oklahoma, population seven thousand, amid pasturelands, stockyards, and fallow cotton fields, far from any ocean.

For a year and a half, he lived and worked in that cavernous hangar, constructing *Idabel*. In the summer, in a tarped-off corner, he slept in a hammock he custom-made out of heavy-duty plastic mesh gleaned from a junkyard (where he also found his prized Scottish cashmere sweater). In the winter, he slept in his car, a 1968 Buick Electra. "I towed *Idabel* with that car," he says, which so alarmed the Oklahomans that they would roll down their windows and offer to lend him their trucks. When Stanley finished the sub and deduced it had a crush depth of about three thousand feet, he had it towed to Roatán, where he was hell-bent on voyaging into "the land of perpetual darkness," as he puts it, the habitat of creatures whose existence predates that of the dinosaurs. Down there, "there are limestone arches," he says, "covered in thousands of sea lilies ... It's otherworldly."

~

*Idabel* is suspended over a rectangular hole cut into Stanley's wooden dock by a polyethylene rope and grappling hook beneath a corrugated awning onto which Stanley has written GO DEEPER in bright red paint. The sub is roughly the shape of a tiny helicopter with a bulb on top—thirteen feet long, eight feet tall, six feet wide. From the side, it resembles some robot snowman wearing a hoop skirt. Stanley mixed two different paints together to get Idabel's yellow the shade he wanted—canary meets hot dog mustard. He wants the sub to look garish so, should he become trapped at depth between boulders, a rescue ship could spot it. But since no rescue ship patrols the sea here at two thousand feet, such a vessel belongs more to the realm of myth than to ours. And if somehow *Idabel* dislodged from such a position, it would likely sustain damage that would not allow it to rise. One engineer who tank-tests subs for the Navy believes it would instead be pulled even deeper, reaching its crush depth, where "it would collapse, irreversibly and catastrophically." Still, who knows? Stanley thinks miracles are happening all around us; we just need to retrain ourselves to see them. It does rain fish here, after all.

"[Some industry] people say things about it being asinine," Stanley says, "for anybody to be operating a manned sub without communications. Everybody knows who they're referring to. There's nobody else doing that. But it's calculated. I don't expect anybody around here to be able to rescue me, so what's the point of communicating? It's adding another layer of complication. I don't want to use 'em."

He relays a story about *Idabel* getting caught on a nylon line when trying to salvage a shipwreck. The sub hiccuped in the middle of the ocean, and the line fluttered, some umbilical tether, all the way to the sunken ship. Stanley took shallow breaths. There was enough air inside the sub for him to survive for three days. Barracuda peered in through the porthole. As the hours passed, the sea grew dark, then darker. Though he was finally able to liberate *Idabel* eight hours later, he admits, "The sub could've hung there for months, if not years, until some leak finally managed to flood it and it sank to the bottom, all the while having us inside, heavy into the process of decomposing."

To get into the sub, I must drop into it, bracing my arms on either side of the hatch. The opening is just wide enough for me to fit, tight as those culverts my mother warned me about playing in as a kid on the outskirts of Chicago—perpetuating the suburban legend about some curious boy who crept inside, got tangled in trash, and died in the darkness, the rats rendering him to his skeleton, which washed out a year later. My mother would not approve of this.

Wriggling through the entry chute, I lower myself through Stanley's captain's chamber, which is lit up red like old New Orleans, a series of dials and levers and cables stuck to the hull's interior—some welded, some seemingly attached with Velcro and superglue. His captain's chair is a tattered bicycle seat. A ring of portholes haloes Stanley, and he looks like a cartoon that's been thumped on the head, encircled by daffy birds. I emerge into the bottommost sphere and sit on a small bench.

A plastic fan whirs over my left shoulder, and I have to hunch to keep from knocking it off the hull. In front of me is that pristine Plexiglas viewport—four inches thick and nineteen inches in diameter. There's another porthole at my socked feet, which rest on a burlap bag filled with lead shot—a weight that I will later pass up to Stanley in order to balance *Idabel* in the deep. *Idabel* has the feel of a giant squat thermometer, and I'm curved into the bulb at the bottom. The space seems hardly larger than a laundry dryer, and I press my nose to the viewport and breathe and wait and hold my breath. It feels airless and humid. But I have brought my long-sleeved shirt in a plastic grocery bag, as Stanley says at depth the hull can become ice-cold and wet with condensation. He pulls the heavy hatch door closed and seals it. My heart races, and I sweat and bite my lip and I have to say something to him.

"It's like we're in a vacuum," I say. My hands shake. If the sub fails, we could suffocate. If we attempt escape, we could die from pulmonary hemorrhages. We could be knocked unconscious by the pressure, or succumb to arterial gas embolisms, or fall to immersion-induced pulmonary edema. We could certainly drown.

Through the porthole between my feet, the seagrass waves and skinny electric-blue fish emerge, then disappear into the blades. Stanley is too busy potchke-ing with the dials to answer me. We descend from the crane, and the ocean laps up over the viewport, and soon we're taxiing toward the subsea canyon where we will drop. Over the motors' drone, Stanley speaks of the deep sea as a stable zone, oddly suspended in time, as if a burial ground and a museum—interring vessels, the skeletons of long-extinct creatures we still don't know ever existed; the remains of fish and cephalopods, invertebrates and people. This is him, tour-guiding.

"I had a tenant suicide last May," Stanley says over the motors' rumble. "He did it in the unit I used to live in. I'm the one who cut him down. I think Covid pushed him over the edge. He basically haunted my house. Kujo detected him and freaked out. Alerted me to the presence of his ghost. We buried him at sea. I've been part of other burials at sea. I almost cut his head off to save his skull too, mount it on the house, but the authorities, and I heard he had like four kids, so . . .

"I think more people should be buried at sea," he continues. "It's the greenest. A lot of people would be pleased with the idea of their body being eaten . . . given back to nature in the most direct way possible, and benefiting a large wild animal that way. There's no other socially acceptable way to feed large wild animals your remains. People are like, *Don't feed the bears, don't feed the crocodiles*. Sixgill sharks: it's the least we can do for them. During the day, they never come up above a thousand feet. Perfect scenario. Like, a human to them is two bites."

Attached to the steel encasing us is no phone, no radio, nothing to cry *Mayday* into. Any distress we experience at depth will be known to us alone. "Here we go," Stanley says. He throws a lever upward, and the air rushes from the ballast tanks. We go under, the depth-measuring dial to my left jumping to life.

# 26

IN SPITE OF the urgent efforts made to locate and rescue the missing OceanGate *Titan* submersible, which had intended to tour the wreckage of *Titanic* on June 18, 2023, many amateur submersible builders were pessimistic from the start. Speaking about *Titan*'s then-recent disappearance, Karl Stanley told me, "I am 99 percent sure they imploded." The p-subbers hadn't been this active online since the Wall/Madsen case. According to posts by the PSUBS group on Facebook soon after *Titan* disappeared, "the currents there are a bugger. They could easily [be] drifting out to sea."

"From what I understand, the sub must be opened from the outside," another said.

"I heard this which is absolutely crazy. If they can't get out [they] can't put up a radar reflector, if they have one of course which they probably don't. It does sound like a poorly thought-out system," responded another. "I'm afraid we will never hear [from] them again."

Many wondered why the vessel, which had made such tours available to those able to pay the $250,000 price tag for the experience, wasn't more secure. Some argued that should something go wrong at such a depth, a rescue attempt would most likely be futile anyhow. Stockton Rush, OceanGate's CEO (and one of the five who was on board *Titan*), claimed one year prior that there is "a limit" to safety. "At some point," he said, "safety just is pure waste. I mean, if you want to be safe, don't get out of bed." OceanGate's website professed that governmental regulatory processes for such a submersible could be "anathema to innovation." Rush believed that the safety regulations imposed on submersibles that wished to be certified were "obscene."

Do many in the community still downplay the dangers of their hobby? I don't know, but I do know that they often seem to get lost in fabricating the exhaustive safety precautions they must take to mitigate, but never eradicate, the encompassing danger. Because they're doing all of these things, and taking all of these precautions, and lending their minds and bodies to inventing and engineering and then physically welding and sealing and gluing and fire-treating the parts for these safety measures, that's what many of them fixate on—the safety precautions. That's what takes up their time and energy, and sometimes, if only rhetorically, many seem to confuse that for actual, encompassing safety, actual encompassing security, when in reality the encompassing thing is the danger—in the literal form of the ocean—the fickleness and power of the deep.

The submersible community is a small one, and Karl Stanley was a friend of Stockton Rush's. In April 2019, Stanley himself was part of a four-person crew (Rush among them) who dove on *Titan* to a record-breaking, *Titanic*-level depth of 12,336 feet off the Bahamian coast.

"I think the main reason I got this opportunity to dive," Stanley wrote in an email to Rush on April 18, 2019, the day after the descent, "might have the most to do with if there was such a thing as 'expert in risk assessment in one-off, uncertifiable deep sea manned vehicles' my resume is hard to beat, and you know we are like-minded when it comes to judging how far things can safely be pushed . . . I believe in what you are doing. I think the use of composites is long overdue for MUVs [manned underwater vehicles]."

But Stanley also rang an alarm bell about the safety of the sub. It took some impassioned arm-twisting to get Rush to heed his warning.

"The sounds we observed yesterday," he continued, "sounded like a flaw/defect in one area being acted on by the tremendous pressures and being crushed/damaged. From the intensity of the sounds [and] the fact that they never totally stopped at depth . . . [it] would indicate that there is an area of the hull that is breaking down/getting spongy." In the remainder of the email, Stanley implored Rush to delay any further diving and to conduct more research on the materials he was using.

"I don't think if you push forward with dives to the Titanic this season," Stanley wrote, "it will be succumbing to financial pressures, I think it will be succumbing to pressures of your own creation in some part dictated by ego to do what people said couldn't be done."

"Keep your opinions to yourself," Rush responded. "[You fundamentally misunderstood] your role while visiting us in the Bahamas. I value your experience and advice on many things, but not on assessment of carbon fiber pressure hulls . . . I hope you of all people will think twice before expressing opinions on subjects in which you are not fully versed."

"How I like to make decisions is by considering worst-case scenarios," Stanley pressed. "The worst-case scenario of delaying diving until you have identified the defect making all that noise is some disappointed customers and financial woes. The worst-case scenario of pushing ahead and not listening to the hull yelling at you involves Patrick Lahey [CEO of Triton Submarines, which fabricates subs for the billionaire circuit] and some Russian oligarch tooling around a Russian nesting dolls version of a wreck site in a made for TV special, telling his version of how things went wrong. I hope you see option B as unacceptable as I do."

Due to this alarm bell, and to Rush's apparent resistance to addressing it, Joel Perry, another crewmember on the dive and once the director of media and marketing for OceanGate, abruptly quit the company. Others who spoke out against Rush's "pushing forward" were fired. A wrongful termination lawsuit was settled out of court.

Four years later, with regard to the widespread reporting on June 21 and 22, 2023, of sonar-detected "noises" in *Titan*'s recent dive vicinity, Stanley didn't believe it indicated any reason to be optimistic. It may be that the news cycle was deliberately extending the story, sowing false hope, though I certainly understand the need for even false hope in the face of tragedy sometimes. On the evening of June 21, when I followed up with Stanley on the fate of the sub after the reports of the "noises" came to light, his response was succinct and final. "They imploded," he said. In the late afternoon of June 22, this implosion was finally confirmed after the Coast Guard found a debris field in the area.

"It was the carbon fiber tube that failed," Stanley said. "Stockton was fixated on a larger number of people getting to the depth of the

*Titanic* and every other decision was secondary. [He] put too much faith in carbon fiber as a material, financial pressures from investors, and high-net-worth clients pushed his timetable. These are the root causes... This was only dive fourteen to the *Titanic*. I suggested to Stockton he should have at least fifty dives under his belt before going commercial... He should have listened."

Many in the personal submersible industry became worried about how the disaster would affect their passion. "We will need to keep on our toes regarding any potential impact that may be imposed on us," Jon Wallace, the PSUBS founder, warned, to which Shanee Stopnitzky responded, "We were given permission to dive very easily in the past, and I can't imagine that being the case after this incident."

Stanley, though, felt that his price point would prove attractive to those tourists still interested in diving on *Idabel*, and could potentially boost his business. "The price point of $250k is in a lot of minds. Twenty-five years of experience and $1200 an hour is looking mighty reasonable."

One member of the PSUBS community, trying to draw parallels between the obsessions of those within it, raised the issues of luck and inevitability, telling Stanley via Facebook, "You had the same kind of doubters [that Rush had] back when you started. Now that you've made over 2,500 safe dives, you have earned the respect of all and [are] looked to for answers, but one tiny critical mistake in 25 years, and you would be judged by many to be a failure. Taking risks means failures are inevitable—catching them before they do damage is the only way to stay alive."

The post quickly generated outrage among the community, and resulted in a pile-on. Shanee Stopnitzky, for instance, was the first to respond: "████████████, if your submersible catastrophically failed, you cut corners implicitly. Bad luck would be getting caught in a ghost net or something... Engineering failures mean cut corners... period."

The original poster tried to distinguish between "knowingly cutting corners" and "making an error you didn't catch," and acknowledged that their hobby is, by definition, riddled with "incredibly high risk, full stop," but his peers weren't having it.

"Stockton was a renegade maniac and he persuaded other people to engage with his maniac dreams," Stopnitzky said. "That is not representative of the sub community-at-large."

Stanley succinctly shut down the conversation. "[Rush] seriously fucked up."

≈

"The OceanGate disaster..." Scott Waters, captain of the *Pisces VI* sub, sighed when speaking with me a few days after it had been announced that everyone on board *Titan* had perished. "This has been known by the entire community now for well over five years," meaning that they were all simply waiting for this to happen, based on Rush's flouting of the safety regulations set forth by the Marine Technology Society (MTS), which governs the professional submarine community (though Rush's enterprise did not define as "professional"). "We urged him," Waters stressed. "We were all concerned about the carbon fiber, but another major requirement violation was they even lacked the ability to escape [*Titan*] when on the surface. The OceanGate team just chose to ignore it... *Titanic*, *Titan*. It's unfortunate."

"Basically, everyone in the whole industry was aware of this [recklessness] happening, and was deeply concerned," Stopnitzky confirmed. "Karl wrote an email to Stockton that was saying, you know, *You're gonna die*... So, Scott Waters asked me to put out a press piece about OceanGate so we could warn the public, because Stockton was a charismatic persuasive type. So, me and Scott were toying with this idea [of releasing the press piece], but we just kind of dropped the ball... And, of course, as soon as the sub went missing, the whole sub industry knew exactly what happened. And it was just horrible to think of that outcome for innocents who were probably misled. I'm totally fine with Stockton dying on that thing. He made it."

"History has a tendency of repeating itself," Waters said. "When you have someone just completely and totally ignoring the rules that are set in place to keep you safe, chances are something will eventually happen, and it did."

For a handful of intense days, the world was fixated, as the disaster seemed to tap into and fuse many of our greatest fearful hits—

claustrophobia, asphyxiation, drowning, being seduced by the deceptive charisma of a narrowly brilliant sociopath, the nightmare of being lost and spinning out of control into a world that isn't quite our own, extreme, extreme isolation . . .

"There hasn't been a moment that's reignited people's inherent fascination with submarines until this," Stopnitzky said. "Imagine if we could activate that excitement and fascination without a fucking horrific tragedy."

The news cycle, once again conjuring its entertainment industry luridness, stretched the story out, fully aware that experts like Stanley, Waters, and Stopnitzky, and those laboring for the Coast Guard, believed there was no hope of rescue. Stopnitzky's take on the coverage was hauntingly similar to my own on that of the murder of Kim Wall.

"It made me realize overall just how little space there is for nuanced representations of reality in media," Stopnitzky said. "The media response," she told me, "was a horrifying and deeply emotional experience for me. I wasn't so sad about the incident. What made me sad in like an existential way was watching the mainstream media just peddle fucking nonsense, and then watching the entire world have a fucking opinion when they know nothing."

The ways in which the story was reported ingeniously exploited our collective anxiety by stitching it, almost cinematically, to a time clock—a suspense-inducing breathable-air countdown clock. Of course, we all became so keenly aware of the passage of time, and the hours ticking away in our own lives too. It's downright stress-inducing—seventy-two hours left, forty-eight hours left, twenty-four hours left, five, four, three, two, one . . . In this, even if implicitly, we were also made keenly aware of how much closer we are to our deaths. We were implicated.

"It was hard to watch," Waters said.

"I'm heartbroken for all the loved ones of my fellow deep-sea travelers," Stopnitzky said, "and hope they get some solace that as far as dying in an accident goes, it doesn't get any better than instantly becoming one with the deep." And while I think Stopnitzky's heart may be in the right place, and while I appreciate her romantic tendencies in most cases, it feels a little presumptuous, and like thin salve in this case. Also: How does she know that? In this case, becoming one with the deep was the result of approximately six thousand pounds of pressure per square inch exerting itself on the human body (or about eighteen million pounds of pressure exerted on the body entire), which, in under a millisecond, rendered the bodies of the five passengers to gel, which was then sucked out into the ocean through the ruptures of the imploded sub. We are here, then we are gone. Dispersed into the brine, as if at the striking of a delete button.

"It's really unfortunate," Waters repeated. "Everybody tried to stop [Rush]. But you can't enforce the rules when you're in international waters."

Stanley, of course, knew all of this when he boarded *Titan* back in 2019. "Stockton warned us ahead of time," he said, "that the submarine had made many loud noises." One may wonder then why Stanley committed himself to the dive, went through with lowering himself into the vessel. Could he simply not resist the opportunity to sink to such a depth? Was the profundity of 12,336 feet too attractive to resist, in spite of the danger? *We are like-minded when it comes to judging how far things can safely be pushed. I believe in what you are doing.*

I recall how, years ago, Stanley tested his own limits, pushing his first sub, *C-Bug,* beyond its operating depth rating, risking implosion, permanently deforming its hull, narrowly escaping with his life. How did Stanley define *safely* in the moment *Titan* dropped beneath the surface and started making those awful sounds? How did he define *like-minded*? For Stanley, when does *far* become *too far*? Similar questions kept me awake at night as I anticipated—sometimes dreaded—my own dive down to two thousand feet with Stanley.

In spite of Rush's warnings, and the alarm bells that Stanley himself would ring the very next day about *Titan*'s condition in 2019, Stanley still felt that "It was not anywhere close to catastrophic ... I didn't feel that our life was really in grave danger at that point ... I also considered Stockton a friend. I admired his adventurous spirit and willingness to push limits." And I can't help but think how this seemingly admirable penchant for limit-pushing wends its way through the amateur submersible community, slowly gathering tragedy. I can't help but remember how Stanley told me that he felt that Peter Madsen murdered Kim Wall "for the thrill. To see if he could do it." To push his limits.

But Stanley, as if navigating the conflicting intricacies of his own psyche, also called himself Rush's "guinea pig" on the dive, and yet, even knowing he was such a guinea pig, he decided to go through with it. I wonder how he distinguished between *danger* and *grave danger*. Later, when *The Onion* posted an article with the acidic headline "Critics Say Submersible Should've Been Tested with Poorer Passengers First," Stanley reposted it and commented, "IT WAS." And yet Stanley is the sort of personality who boarded *Titan* anyway, as it dropped to record-breaking depth, making all that noise.

"I think there is almost no limit to risks people are willing to accept for such an adventure, as evidenced by this year's season on Everest," Stanley said. "[But] the time frame that [Rush] did this all on was beyond any common sense. He was willing to take people out to the middle of the North Atlantic when he had done only four deep dives and probably not even one without a major system failure. I am beginning to wonder if he was suicidal."

And I am beginning to wonder if Stanley may have such tendencies too, as, by his own assessment, his boarding of *Titan* in 2019 was also "beyond any common sense."

"As far as going down [on that dive on *Titan*] in 2019," knowing what Stanley had known, Waters told me, "there's no chance in hell I would ever have done that. And I'm very fine with taking calculated risks . . . But with Stockton's design, it's completely crazy."

Stopnitzky agreed. "I definitely, for sure, would never have gotten onto that sub," she said. "My friends think I'm incredibly reckless and crazy and do all kinds of dangerous things all the time, but in my mind, I'm not willing to tolerate low-to-medium probability with catastrophic outcomes. It's completely unacceptable to me. I drive like a grandma. I mean, I also make explosives . . . I do think that desire to be first and that conquesting sensibility is usually male . . . I don't have an attachment to that kind of pioneering reward. So, no. Even going into Karl's [sub] was pretty tough. I was like, *I might die and I'm okay with it in this case,* because I wanted any experience of the deep sea. But with something like *Titan*, no. Dying is not worth that experience."

≈

"There are three ways to get into ocean exploration," Stopnitzky told me. "One is as a scientist. And not just any scientist, but highly regarded and educated, PhD-plus, and even that is pretty rare that you'll be going on a sub dive. One is as an engineer—someone who works in submarine operations. And still, there's this enormous degree of competition for a space on a submarine. And then there are the people who buy their way in. And those people generally control the narrative. They have the money to control the communication around these events. They're usually driven in part by ego, and so the story out in the public and in popular imagination is what [the wealthy] get out of it. The really human stories are not being told—the psycho-emotional experience. The billionaires' stories are not representative."

Some who own mega-yachts are commissioning the construction of their own personal submersibles, which some tether to the aft of their vessels—just another toy leashed to a larger toy. Others hire "yacht integration" companies to design storage holds for these subs, so they can be seamlessly and efficiently incorporated into the yacht's design.

"The rich want the sea to themselves," Stanley told me, and bemoaned companies like Florida's Triton Submarines, who "make subs for billionaires," or, as they're called on Triton's website, *discerning individuals.* Triton advertises that many of its subs are *completely devoid of compromise* and designed not just for yacht use, but for *super yacht use,* and for *owners of Megayachts, Gigayachts or one of the new breed of Expedition/Explorer yachts* looking for *the ultimate luxury experience,* and *the privacy, amenities, comfort and luxury usually experienced in a private jet.* Owners and their guests can enjoy designer lounge seating and minibar while gliding amidst a school of magnificent sharks.

"A lot of people in the industry—at least I hope so," Stanley said, "are saddened by the way it's gone. Submersible [exploration and research] used to be publicly funded, driven by science, and now it's just the uber-rich. So that seems to be the future. Making toys for the uber-rich... There was this billionaire who commissioned a sub that went to the bottom of the Marianas Trench [seven miles below sea level]: when he first approached Triton Subs, he wanted it to be for one person and he wanted to bring his little dog. That's where the market's at. It's like this billionaire, this billionaire. It's like little playing cards to them. Complete toys for people to bring their dogs to the bottom of the ocean..."

Of course, not everyone who inherits considerable wealth grows up to be a megalomaniacal, petulant man-child willing to unapologetically sacrifice the safety of other people's bodies, commodifying them to satisfy his own "innovations," and then, if something goes awry, depending on the overarching system of like-minded men-children to control the resulting narrative, absolve their demographic of any substantive blame, and maintain the status quo wherein the well-being of the masses takes a backseat to their need to dictate the machinations of our reality in ways that continue to benefit themselves and perpetuate similar half-baked "innovations." (*Inhale, exhale...*)

Like Peter Madsen et al., Richard Stockton Rush III (who claimed to be a descendant of two pioneering men who signed the Declaration of Independence, and whose widow, Wendy Rush—OceanGate's former communications director—is a descendant of a couple who died on *Titanic*), flush with his family's oil fortune, was obsessed with planting his little flag both in outer space and in the deep sea. According to a former business partner, Rush was a "frustrated astronaut" who once dreamed of "making humanity a multiplanet species." When his rocket ventures fizzled, he turned his eye to the deep sea and founded Ocean-Gate, keen on establishing himself as the Elon Musk of the ocean, and marketing himself to "wealthy tourists" thirsty for "extreme travel." One wealthy customer described Rush as "can-do."

Rush dubbed his own seemingly addictive attraction to the sea "the deep disease," and he was hell-bent on infecting others and getting even richer. He wasn't content with being a passive passenger, and he was even dissatisfied with being any old ordinary captain. He wanted to be the biggest captain. The best captain. "I wanted to be Captain

Kirk on the *Enterprise*," he gushed, "I wanted to *explore*." Like Madsen, Rush had often been described as exhibiting a "child-like verve." "The future of mankind is underwater," Rush said. "We will have a base underwater . . . If we trash this planet, the best lifeboat for mankind is underwater," and one may wonder if by *we*, he meant the uber-wealthy, and if by *mankind*, he also meant the uber-wealthy. If he intended for the world to be inherited by his own Mega-Giga crowd.

"We figured going undersea was as close as you could get to space without leaving Earth," Rush's former partner, a so-called serial entrepreneur and angel investor, said.

And in spite of the alarm bells rung by the industry, and in spite of the cold, hard fact that they were right in ringing them, Rush's former partner still made excuses, post-disaster, for Rush's insistence to move forward with the doomed dive, claiming that OceanGate's critics were being unfair since they "didn't work at OceanGate, they weren't part of the technology development program, they certainly weren't part of the testing program . . . Regardless," he said, "everyone has their own opinion." One may wonder what "opinion" has to do with any of this in the aftermath of the fact of *Titan*'s implosion.

≈

In the days before that fact hardened into itself, I had desperately hoped that *Titan* would be found intact, and the five people on board would be rescued. But I was, of course, afraid that they would not be, and I'm still wondering if this fear—like many others—is not one that should be overcome. Do we really need to sink to such depths to slake our wonder, prove our wealth or untouchability, actual or metaphorical? What's it going to take to dismantle these seemingly rock-solid systems of control that incubate and protect the narratives of those with spoiled Captain Kirk fantasies—and the means to manifest them—at the expense of the rest of us? Or are we too far gone? Have our hulls already begun to degrade? Are we just making noise? Is anyone listening? Are we already imploding? Have we been for some time now?

≈

Twelve days after *Titan* imploded, following up on our initial correspondence about the tragedy, Stanley reached out to me via WhatsApp, claiming that Stockton Rush knew exactly what was going to happen on *Titan*'s final, fateful dive, and that Rush intended all along to build what Stanley called "a mousetrap for billionaires." "He was in fact a good engineer," Stanley said, "and that machine did exactly what he designed it to do . . . He set a new standard for going out with a bang."

Stanley claimed that Rush named his doomed sub after the fictional British ocean liner *Titan*, from the 1898 novella *Futility*, written by Morgan Robertson. In the eerily prescient novella, the fictional *Titan* (initially believed by the book's characters to be unsinkable) sank in the North Atlantic after striking an iceberg. Fourteen years later, in 1912, after the actual, and hauntingly similar, *Titanic* disaster, the novella was reissued, its title changed from *Futility* to *The Wreck of the Titan*.

Robertson, who claimed to have invented the periscope, was accused of clairvoyance, which made him uncomfortable. "[Rush's] ego was so big," Stanley wrote to me, referencing the resonance of *Futility*, "he was willing to die and kill to be the pivotal character of this story. He wanted to go [die] at the wreck [of *Titanic*]. The more high-profile, the better. He didn't just murder four wealthy people and get paid a cool mill to do it—they are allll part of the *Titanic* mythology now." According to Stanley, in order to extend, and to be a part of this larger story, Rush needed to compel more than just his own death, and he needed to knowingly fabricate a "futile" vessel, costumed in a titanic name, as his murder weapon.

And here Stanley quoted one of his favorite bumper-sticker slogans—something Rush apparently used to say—"Live a good story." He told me that most everyone who worked for OceanGate knew that this was going to happen and that "all of them are shitting their pants right now. But, [Rush] being a madman who manipulated [them] is the best narrative for them."

"Let me get this straight," I wrote to him. "You believe that Rush knew the sub was going to implode—actually intended it to—and that he knowingly killed everyone onboard, himself included, to best become part of the ongoing *Titanic* mythology?"

"I know this is what happened," Stanley said.

"Do you feel lucky to be here?" I asked. "Do you now regret taking him up on that 2019 invitation?"

"No and no," Stanley said. "I got to go to 12,500 feet. And lived a good story."

"And," he said, "[I] didn't pay 250K."

## 27

As the Copenhagen police, led by officer Jens Møller Jensen, conducted their investigation, beginning with a thorough search of *Nautilus,* which Tonni Andersen and his *Vina* crew had recently raised, the days seemed to pass in a blur, unstuck from time. Words like *slow* and *fast* didn't apply, the ornaments of the world having gone matte. There was a numbness to going about one's day, as if all sensory stimuli hung at a distance—untouchable, unhearable, as if behind a curtain of Visqueen, as if some kind of surface glimpsed from underwater, if the water weren't water at all, but gel. As the police searched for evidence, and determined finally on Sunday, August 13, that *Nautilus* appeared to have been deliberately sunk, Madsen remained in custody. Under Danish law, he could be held in this way until September 5. Any letters mailed to him at the prison were read first by the police. Outside of meetings with Engmark, he was allowed only one hour of supervised visits per week, during which he was forbidden from discussing matters of the case.

Kim's family drove from Trelleborg, Sweden, to Copenhagen to collect her belongings, and some friends of hers and Ole's helped them load their car, crying together. The wheels of Kim's old black suitcase were worn down and wobbly after crossing the airports and streets of so many cities in so many countries. Inside it were her favorite pens—the transparent ones with Japanese characters on them—and her silver laptop in the case she bought in China. The police invited the Walls to see *Nautilus,* but why would they want to see such a thing, and barrel through a mosh pit of reporters to get to it? They returned home, and when the police visited them there, the Walls handed over

Kim's hairbrush and toothbrush. People were searching for her body, the police assured them, and the Walls told the police to look for the orange sweater she had been wearing; it was so bright.

The days dragged themselves through those who endured them. "We get up every morning. Take a shower. Walk the dog," Wall's parents said. "But nothing feels real . . . Every walk along the beach is a search for flotsam." The Baltic Sea took on a new air of menace. Reporters rang and rang their doorbell. Every time, they jumped at the sound.

It was on Monday, August 21, nine days after Madsen's initial court testimony, that the bicyclist came upon the headless, limbless human body while riding along the shore of Amager, southwest of Refshaleøen. The body also bore several stab wounds and was cinched with a strap securing metal weights. The asphalt bike path was blocked off, and a white sheet laid over the remains. The weather was unbearably beautiful—sunny, 68 degrees. Two days later, on August 23, at 2:05 in the morning—in the middle of the cool, clear night—having used DNA from her hairbrush and toothbrush, the investigators confirmed that the remains were Kim's. They also found traces of blood on board *Nautilus*. For the Walls, waves of despair crashed against waves of love. They told their favorite Kim stories—the one about the rainy Fourth of July, on an isolated beach, when Kim couldn't stop wondering about the story of a lone man sitting beneath a giant umbrella; how, when she was on assignment overseas, she wrung her hands after her vegan host family ate a piece of nonvegan cake she left in the refrigerator, and loved it so much, they asked her for the recipe. The joys of cake. The sweetness of cake. The ethics of it.

With a German shepherd named Cross and a Labrador named Ace, the police scoured the shoreline for the remainder of Kim Wall. The dogs were specially trained "to pick up the movement of dead people in water." Investigators soon boarded police boats and cruised the same route as *Nautilus* did on its final dive. The dogs sat at the boats' bows like hood ornaments, barking into the wind, snarling at the water.

When confronted with new evidence, Madsen changed his story, then changed it again, then changed it again. Somehow, people had gotten ahold of the Walls' contact information, and, on top of everything, they had to delete messages from internet trolls. Engmark's private cell phone number got leaked, and she received a barrage of threatening text messages. Engmark was rattled, but tried to take comfort in the notion that she was offering a service that defended the rights of the accused, a right that was a pillar of any just society, she told herself.

On Sunday, September 3, two days before his remand expired, compelling the court to officially decide if he should remain in prison, Madsen requested, through Engmark, that it be an open-door hearing, allowing access to as many members of the media as possible.

≈

Grizzled, world-weary prosecuting attorney Jakob Buch-Jepsen wiped the sleep from his eyes with his sleeve, wiped his nose with a square of toilet paper, and began shaving. The deep parentheses at the sides of his mouth made it difficult, but he was practiced in this difficulty. He was practiced in many difficulties. His close-cropped sandy hair, perfectly faded sideburns, and reddish complexion lent him the look of a spent Daniel Craig, but leaner in the jaw, and puffier under the eyes, like an ex-boxer—one who went about 50/50.

Once clad in his typical getup—black suit, blue tie, white shirt, long black overcoat—he appeared as if he couldn't suffer fools any less. He appeared as if, should he conjure the energy, he would take their eyes out. If he were a dog, one would only reluctantly, and with an anxious heart, hold out one's fingers for it to get a whiff and muster acceptance. He appeared as if he were holding a decade's worth of growls inside of him, and only a saintly restraint had kept him from baring his teeth. But, as with many saints, maybe that restraint was eating him alive.

Buch-Jepsen's phone rang. He shook the water, stubble, flakes of skin from his razor. He had just finished successfully prosecuting a triple-murder case. For years, he had had a close working relationship with the Copenhagen police and had been their go-to solicitor for what they deemed "dangerous crimes." He was overworked, and he swallowed hard, his Adam's apple jumping as he agreed to lead the prosecution of Peter Madsen.

He took his clothes out of the closet. Controlled his breathing as he buttoned his shirt, cinched the knot of his tie close and tight, threw on his overcoat as if a cape. He headed to his office to do preliminary research, make a few phone calls. The office building was sterile and white and across the street from an undeveloped field, which was fronted by a tall chain-link fence topped with barbed wire. Buch-Jepsen's office building shared space with EPI: Top Quality Flooring Systems and Coatings, their slogan: Design Your Life! And it shared space with EcoLab Nalco, a wastewater recycling firm, and Best One online marketing, and Tiponi social work, and Neo Coating epoxy, and Valby Judo Club, and Ny Liv spa, specializing in lymphatic massage, coconut oil massage, hot chocolate butter massage, bamboo stick massage, and full body waxes.

Across the street, on the other side of the vacant lot, stood another industrial building, housing the offices of Service Forbundet, a trade union representing white-collar workers, and BCD corporate travel agency. Occasionally the employees would run into one another at The Organic Boho, the vegan joint up the street, or at Samvær, a restaurant with an identity crisis, serving Italian, Indian, Thai, and Spanish dishes (though apparently their spaghetti and meatballs, advertised on the menu next to that famed image from *Lady and the Tramp,* is to die for).

Buch-Jepsen's office didn't have exterior signage, but lurked in the depths of what was simply labeled OFFICEHOTEL, a generic bank of spaces for rent—some full, some empty. Many employees bicycled to work, the racks outside overflowing with overlapping tires and handlebars. He walked in, rolling his briefcase past spaces peddling polyurethane laminate, and spaces wherein kids threw other kids over their shoulders onto padded mats, and spaces where people were oohing and aahing as hot chocolate cascaded over their bare backs. He kept his head down, eyes down. From the judo club, and from the waxing rooms, there occasionally would be screams. Buch-Jepsen unlocked his office door. On the walls hung pictures of the ice fields of Greenland. On the shelves stood figurines of polar bears and musk ox. He put on a pot of coffee and rounded his desk, sat in the tawny cushioned chair. A polar bear peered over his left shoulder.

On Tuesday, September 5, the day of Madsen's detention hearing, the press lined up at the courthouse door at sunrise. They came from all over Scandinavia and beyond. The journalist at the front of the line was from France.

Buch-Jepsen drove here from his office through the industrial park. He parked, pulled his black trolley case stuffed with documents over the cobblestones of the Nytorvet square leading to the courthouse, weaving through the journalists local and foreign, ducking their microphones and tape recorders. He kept his head down, eyes down, and the wheels of his case sputtered like some rickety fairground ride. In his case was an official request to keep the courtroom doors closed, to avoid turning Madsen's trial into even more of a media circus (though Madsen and Engmark continued to ferociously argue that the doors be thrown open). He also had documents requesting a mental health evaluation of Madsen, and a demand that Madsen's computer be turned over to police.

He climbed the sixteen white steps and touched the green doors, held their cold, cigar-shaped black metal handles. He made a beeline for courtroom #60, the building's largest. The press closed in behind him like a wall of water.

Shockingly, the court denied Buch-Jepsen's plea for a closed-door hearing and decided to grant Madsen's and Engmark's request. The doors were thrown open to the media. They jockeyed for position and rushed inside, desperate to secure one of the too-few chairs. An adjacent courtroom had to be opened up to accommodate all of them, and an audio feed of the hearing was piped in. Engmark positioned herself at the left side of the courtroom, Buch-Jepsen at the right, standing behind a tower of green binders overflowing with dog-eared papers. The courthouse bells rang once, twice, and Madsen was brought in, hands cuffed before him, clad in his military-green jumpsuit. The journalists whispered among themselves, and Buch-Jepsen looked appropriately annoyed that the judge ruled that he had not provided sufficient evidence that the open doors would "decisively harm the investigation." He held on to his pillar of green binders as if to keep himself together. His face bunched at its center, and seemed to darken, as if some unexpected weather had passed over, or a series of blood vessels decided to break at once. He looked stormy and bruised, capable of knocking out the power.

Madsen smugly postured for the journalists and pontificated self-importantly about his own genius, sticking to his latest story. Acting like an awful fool, he spoke of having "quite ordinary loving erotic intercourse" with many women on board *Nautilus*. He spoke of enjoying red lipstick, stilettos, and nylon stockings. He waved his arms in the air as he said this, some terrible marionette. His cheeks went rosy, and his hair flew. The journalists, of course, ate it up, and their subsequent stories bore a sensational ugliness, devoting outsize energy to the flamboyant Madsen and relatively little to Wall, who was cast not as a brilliant journalist who had long successfully navigated hostile spaces and patriarchal systems in her pursuit of stories but simply as a victim.

As Buch-Jepsen's questions persisted, Madsen became impatient, his answers flatter and more clipped. He worried aloud about who was feeding his cats, who was giving them fresh water. He blamed a passing ship for causing the wave that compelled the hatch to strike Kim Wall. When asked why he didn't radio for help, but conducted the "burial at sea," he glibly answered that he "feels very uncomfortable with the presence of bodies in the submarine. I have no desire to have contact with the dead person. I don't want the dead person in my submarine. I was suicidal at the time..."

Buch-Jepsen's jaw twitched, and he shook his head *no*.

Engmark stood to question Madsen. She focused her line of questioning on the sub itself—the immense weight of the hatch—and on the seriousness of Madsen's brief "suicidal psychosis."

"I know that my world is going to the same place as Kim," Madsen said. "It dies. It doesn't come back." His voice took on a manic pace as he recounted his version of the events, at times closing his eyes and seeming to relive them in his head. "Kim was having the time of her life," he said. "Peter is talking a lot. He is happily sharing his dreams with Kim."

He lapsed into an animated but tedious rant about the technical minutiae of his submarine and what it took to build it. Though Engmark let this rant unfurl in all of its ancillary chaos, the judge had to interrupt numerous times, demanding that he "stick to the essentials." Madsen didn't. Instead, he spoke reverently of the movie *Terminator 2* and compared himself—referring to himself in the third person as either Rocket Madsen or Peter Submarine—to the titular character, hinting at his latent desire to become *part* of the machines he built. Madsen expressed that a part of him *was Nautilus,* and *Nautilus* was him. They shared a brain and a vision. And when his sub went down, part of him went down with it. Perhaps as a distancing mechanism, he imposed some odd inhumanity or robotic quality onto Wall, stating that after she died, he knelt next to her body and "smacked her cheeks to try to reboot her." Throughout his testimony, he applied the terms "her" and "she" to both Wall and to *Nautilus,* so at times it was unclear as to whether he was referring to the woman or the submarine.

"What happens from here is no consequence . . . ," Madsen said. "I have no choice but to sink the submarine." He claimed he had lied earlier to protect the Wall family from the unfortunate truth. Madsen's face contradicted itself. His eyes appeared disingenuously sad, and he held them open too wide, overplaying this so-called sadness. But his mouth was twisted into a subtle smirk, coiling up like an inchworm at its ends, and the bottom of one of his yellowed incisors poked through. Engmark nodded, but not uncomplicatedly. She was frowning.

Buch-Jepsen palmed his stack of binders, decided to use the open-door courtroom against his opponents, and asked that a special report ban be lifted and that the press be allowed to remain in the room as the autopsy report of the remains that the bicyclist discovered was read. The judge granted the request, and as Buch-Jepsen, his voice wavering, intoned the details, the courtroom fell vacuum-silent, punctuated only by the stunned deep breaths of the journalists. They looked at one another with sideways glances. They gritted their teeth and they breathed. Madsen, as the report was read, kept his eyes fixed on the tabletop before him. He looked bored, and his hair was undoing itself, growing more disheveled with each horror Buch-Jepsen intoned.

Buch-Jepsen returned the report to his binder tower. He cleared his throat and collected himself, but his hands remained balled, his fingernails digging into his palms. Fragments of memory swam in his head, like floaters. Fading in, fading out, conjuring ghosts. He stared at Madsen, but in his head, images of trees and shards of trees whirled. Buch-Jepsen grew up in the middle of the woods. His father was a state forester, and the family occupied the little farmhouse on the Arresødal manor property on the outskirts of Frederiksværk—sixty kilometers but seemingly half a world away from Copenhagen. By the time the family moved to the property, the Arresødal manor had been inhabited by generals and princes and kings, mercenaries and philanthropists, people who beheaded other people, and people whose life missions were to "alleviate poverty and misery." It had been a convalescent shelter for women who were ill or who had been assaulted. It had been a safe house for the Nazi occupation. It had been a commune, a hospice, a cult headquarters, a prison. Throughout, it was surrounded by such regal sycamores and maples, chestnuts and an opportunistic invasive species lovingly called the "tree of heaven." In Arresødal, heaven invades. In Arresødal, heaven is a noxious weed.

Buch-Jepsen's father was stoic and shunned hugs in favor of firm handshakes. But his mother was warm and loving, and Buch-Jepsen adored how she would light two tall skinny candles when the family ate dinner together at night, and how she would arrange bouquets of flowers she had gathered earlier in the day as the dining table's centerpiece. Buch-Jepsen's father spent his days roaming the Arresødal woods, and sometimes Buch-Jepsen followed behind. Together, they watched the trees. They remembered things about them and wrote them down. Sometimes Buch-Jepsen stayed behind, in the house, watching his mother in shadow hunched wraithlike over her desk, sketching her clothing designs, his two older sisters whispering secrets to one another down the hall. He watched how the light poured in through the window, nested in his mother's hair, as his sisters whispered and whispered. He watched the shadows of the trees, how—if there was a breeze—they animated the light, and his mother looked like a figure in some Renaissance painting.

Buch-Jepsen stared at Madsen, but he saw through him, to the forest on the other side. He saw the birds that roosted in the branches—spirits, perhaps, of all the women who had ever convalesced across time in the stately but sallow rooms of the Arresødal manor. He saw the light turning earthly things into things we misperceive as angelic, saddling all things dead and gone with unrealistic expectations. "The question is," Buch-Jepsen said to the court, "if it is not Peter Madsen who has dismembered Kim Wall, then who is it? Is it a Russian submarine that has been lying in wait? It is not likely . . ." Buch-Jepsen shook his head. He held on to his binders. He sat down, slowly, as if sinking.

Engmark futilely argued for Madsen's release, claiming that there was no evidence to suggest that Madsen was the one who did this. "He pleads not guilty," she defiantly told the press afterward.

Regardless, the court decided that Madsen was to remain in custody while investigators searched his computer. Kim's parents endured the remainder of September by organizing a memorial fund in their daughter's name. They sorted through old pictures of her; through old videos. They used these to cobble together a press release for the Kim Wall Memorial Fund, and they uploaded the footage to the internet. People found it, and people donated money to help burgeoning female and nonbinary journalists carry out their stories. People found it and held ceremonies in Kim's name all over the world, posting videos of themselves holding candles and singing.

As it must, September stubbornly begot October, and on the third of that month, Copenhagen investigators reported that Madsen's computer housed videos of women being tortured and murdered. At a new detention hearing, Madsen—via a video link from his prison cell—claimed that his interns also had access to his computer and that one of them was probably responsible for downloading those videos. Three days later, on the morning of October 6—a Friday—the Copenhagen police found on the ocean floor a plastic bag containing Kim Wall's clothes and a knife. Soon thereafter, they exhumed another plastic bag. In it was Kim Wall's head. Nearby, on the silty bottom of Køge Bay, they also found two human legs, weighted with chunks of metal to keep them from rising to the surface. That day, Kim's parents were in transit to New York to participate in a memorial service held in her honor at Columbia University, Kim's alma mater. They stayed at a friend's apartment in Manhattan. They drank a little wine together. They checked their messages, and when they phoned Jens Møller Jensen, the Copenhagen investigator in charge of the case, it was 11:00 p.m. in New York, 5:00 in the morning in Denmark. The Walls were concerned that they would be waking him up, but when Jensen answered, he sounded alert—as if he had been awake for some time. Among the news that

Jensen shared with them was that Kim's cranium showed no signs of trauma, findings that contradicted Madsen's claim that she died as a result of *Nautilus*'s 150-pound hatch hitting her on the head.

When confronted with the fresh evidence, Madsen, on October 11, became petulant in his cell, insisting that he did nothing wrong and that he would no longer participate in any hearings or cooperate with further police questioning. The next day, retracing *Nautilus*'s path from two months prior, investigators found a saw at the floor of Køge Bay. The Walls returned from New York to Copenhagen on October 17, and on October 30, at the point when they, according to Ingrid, believed it couldn't "get any worse," investigators revealed that Madsen the previous day had changed his story once again. In this new version, Madsen claimed that he was up on deck, and Kim died below, probably of carbon monoxide poisoning. He admitted to dismembering her body so he could handle her "burial at sea" more easily, but he continued to deny murdering her. It seemed as if even Betina Hald Engmark had had enough of this, as on November 1 she herself publicly rejected the "carbon monoxide" explanation and was unwilling or unable to muster an official comment on Madsen's divulgence that he had butchered Kim's body. Two weeks later, on November 14, Madsen's detention hearing was postponed after Engmark announced that her client—for the time being—was willing to stay incarcerated voluntarily.

Kim's parents withstood the days grieving, walking their dog, watching the weather. They knew that these gloomy, stormy November days would make it difficult for police to continue their probes of the sea. "The days are long and, in both senses of the word, dark," Ingrid said. "We can't do anything—just wait. Wait for a phone call we hope will come, a call that tells us that the search is over—that all of Kim has been found."

In Trelleborg, the Walls attended a gala held in Kim's honor, in an old sports hall. About five hundred people showed up, dressed in suits and gowns. They applauded and applauded as home videos tracing Kim's life played on a shuddering screen. On November 22, Danish police brought up a left arm that had been weighted with a metal pipe. A week later, they brought up a right arm. One hundred and eleven days after her murder, all of Kim's remains had been reclaimed from the ocean. The Walls counted down the days to Madsen's detention hearing. They decided not to send out Christmas cards, as they had in years prior. Instead, they sent thank-you notes to all who had supported them during this time; to all who had helped launch the Kim Wall Memorial Fund. On Christmas Eve, Kim's parents and brother gathered at the family home, focused on the fourth chair in the room; the empty chair. They reminisced about better Christmases. They watched Donald Duck cartoons on TV.

Madsen was officially indicted on January 16, 2018, and charged with murder, sexual crimes, desecration of a grave, and crimes against maritime law. The trial was scheduled to begin on March 8, 2018—International Women's Day. The Walls faced and navigated the barrage of stories told about their daughter and her murder in a seemingly endless stream of newspaper articles and columns, radio shows and premature documentaries, their tones ranging from the respectful to the lurid. On February 1, they and Ole traveled to London to participate in a memorial service held by Kim's UK friends. There, they told stories about her travels, her staticky phone calls home when on assignment, about the music she loved, about her phobia of worms. After the service, they adjourned to a hall where Kim's British friends had organized a table full of Swedish cinnamon rolls. They ate them, and told so many stories, and sometimes they laughed, and sometimes they didn't.

Back home, Kim's parents walked their dog three times a day. On garbage day, they took their garbage to the curb. They vacuumed the house. They vacuumed the house again. In the sky, they watched the swans. On the early morning of March 8, they woke by alarm clock. They drove across the Øresund Bridge to the Copenhagen courthouse where over one hundred journalists from sixteen countries had also gathered. Ingrid described the atmosphere as one of "complete hysteria."

With a police escort, they entered the courthouse through a discreet side entrance and made their way past the spiral staircases, the law archives, the crystal candelabras dangling from the arched ceiling. The press—many of whom had been eating as quickly as possible in the courtroom cafeteria—followed the Walls into courtroom #60, wiping crumbs from their faces and lapels. The Walls took their seats in those marked with the laminated note cards: *Reserved for Next of Kin*. Somehow, they missed the moment that Madsen was led in. He seemed to have simply materialized in the chair next to Engmark, wearing a black T-shirt and blue sweatpants, both of which were too big for him.

From Ingrid's vantage, Madsen appeared to be staring down—fixated on his hands or on his knees. He sat about thirty feet away from them. The lawyers again told their stories, and Madsen again told his—the carbon monoxide version—claiming that he was initially embarrassed to admit that Kim had died due to his overlooking a defect in his equipment; claiming that he was unable to carry her body when it was whole up the ladder to the hatch to give her a proper sea burial. He didn't answer a question about why he didn't call for help. The day was rife with unanswered questions, but the Walls were assured that this was typical of the first day (of what was to be twelve days, spread out across six weeks) of a trial to determine guilt and, if applicable, the requisite sentence. They left through the same covert door through which they had entered. They checked their messages. They saw that, earlier that day, the Kim Wall Memorial Fund had reached its goal of amassing $200,000. They went home.

Over the next few weeks, they sometimes followed the trial days from afar, via live blogs; and sometimes they made that trip over the Øresund Bridge, to courtroom #60. Former members of Madsen's crew were questioned; the videos found on Madsen's computer were shown. The Walls wondered if any of this information would allow the judge a fuller picture of what happened on *Nautilus* the night Kim died, so they could deduce a suitable verdict and sentence.

Spring had emerged beautifully from a typical Scandinavian winter—long, cold, and gray—but the air was still cool and it had rained early in the morning of Monday, April 23, 2018. Once again, Kim's family had to drive across the Øresund Bridge for the closing arguments, and, once again, their hearts grew anxious at the sight of it, and at the water beneath, a feeling they described as "discomfort." In the courtroom, Madsen requested that the Walls not be allowed to walk directly past him on their way into and out of the room, as this would "upset him." Buch-Jepsen argued for a life sentence, Engmark for Madsen's release. The judge had forty-eight hours to deliberate. Madsen turned from his chair, faced the Walls. "I'm very sorry," he said. They wondered what exactly he was sorry about. What a person like him may have meant by this. They wondered what the verdict would be. If, in two days, Madsen would walk free, or be incarcerated for life, or something in between.

Kim's family drove back across the bridge, and the next day, Tuesday—in part to distract themselves from the rabidity of the press—they got out of there; flew to New York to participate in a ceremony held by the Overseas Press Club of America, lighting a candle in honor of all the journalists "who have died, been injured, or been kidnapped while performing their work." At the airport, an elderly woman, though she was a stranger, recognized Ingrid while they waited for their luggage to pass through security, and hugged her. They held each other for a long time.

On Wednesday, April 25, 2018, at 1:00 p.m., Peter Madsen was found guilty on all charges and sentenced to life in prison, the harshest punishment under Danish law. He is only one of a handful of people who are serving such a sentence in Denmark. The Wall family received the news in their hotel room at 7:00 a.m. New York time. "Justice has been served," Ingrid Wall would later write, "but so what?"

Like the Walls, neither Engmark nor Buch-Jepsen has greeted sleep in the same way since. Engmark continued to withstand the harassment of anonymous men who had uncovered her contact information, and grew suspicious of the shadows moving outside the windows of her home. By now, so many had threatened to set her house on fire as she slept. So many had used the word *whore*, used the word *destroy*.

Buch-Jepsen described his nights following the trial as "sleepless." He became acutely sound-sensitive. He spoke of a sadness that settled in during the case, a sadness that followed him home and became, in the aftermath, the unfortunately new normal. He installed an alarm in his home and, like Engmark, grew concerned about threats "outside the house." He struggled to maintain "faith, hope, and love" after having been "staring right into the courtyard of hell."

He, like Engmark, seemed to be mystified as to why this was the particular case to affect him so; to be the straw that broke them. Of course, it was awful, but they had worked tough cases before. Perhaps in part it had to do with the visibility of the case, and their own increased visibility. The lights were always on them; the eyes . . . Given the persistence of the press, they had no escape. They were always expected to say something about the horrible details of the case, think something, feel something, offer a fresh perspective, contextualize horror for the rest of us, all while being expected to frame it as an attorney, conjure the faux dispassion of the professional. After all, what they said would be scrutinized, reprinted, rebroadcast, repeated and repeated and repeated.

Following the Madsen case, Buch-Jepsen decided to take an indefinite leave of absence from the prosecutor's office. He had to get out of Copenhagen. He had to get out of Denmark, even, and moved to Sweden, close to the things that brought him joy as a child—trees and water. Still, it seemed as if he was glimpsing the world though a smudged lens. He tried to break through it, seeking peace and quiet through long walks and breathing exercises, and through writing; through maintaining his late mother's traditions. "This thing about having fresh flowers in the home, lighting candles and hugs," Buch-Jepsen said. "For me it is incredibly important not to be afraid to show love.

"[But] how to preserve yourself . . . ?" he wondered. "When you stand there as a prosecutor, you look the accused in the eyes and look for answers as to why they have done what they are accused of. Trying to understand the incomprehensible . . . There are people who are completely driven by needs."

When the long walks and breathing exercises weren't enough, Buch-Jepsen ran. And when running was insufficient, he ran again— longer and harder and faster, as if he too were seeking the solace of some extremity, some ever-increasing depth. He ran and ran, until it seemed as if all he did was run, all he *was* was running, was speed itself, and the world became a blurred space, muddied and timeless and streaked with paint. Still, running here, one couldn't escape the smell of the sea, and its influence; the way in which, at sunset, it affects the clouds, elongates them. So often—*too* often—they resemble giant

U-boats slowly sinking to meet the sea, agitate all of the things, real and mythological, wondrous and horrible, interred within it.

Buch-Jepsen sputtered his lips. "Running is good because you can manage it yourself," he said. "You are not dependent on others."

He stared out his window. The sun was going down, and the clouds were thin orange lassos, roping other clouds. And somewhere out there, not too far, the sea lurked. Buch-Jepsen took his left pinky into the fingers of his right hand and rubbed it. His cuticles were frayed. The clouds were so orange.

"It can be difficult to be happy," he said.

## 28

Herstedvester Prison, in a dreary suburb fifteen kilometers outside of Copenhagen, is a gray and beige asterisk, some prostrate starfish crucified on a sharp square expanse of green lawn, surrounded by a tall brick wall. In one of the starfish's sorry legs, Peter Madsen now resides, and will reside, most likely, for the rest of his life. To get to the prison, one must drive up a skinny, oil-stained road, wide enough for only one car. If there's oncoming traffic, one has to jump the curb, pull onto the grass. On the way to the prison, the grass on the right is well manicured and fronts a beautiful stand of wispy Danish beech trees. On the way from prison, the grass on the right is an unkempt tangle of kudzu. One navigates a corridor of tall green fencing topped with razor wire, bright orange construction cones lining either side. There's something oddly ceremonial about this entrance, as if one is crossing the threshold to some wan and anemic Oz. One may rightfully expect, as I do, to traverse a sallow moat, a slate drawbridge.

The encompassing town of Albertslund is a planned community, and the low-rise housing units are uniformly built, beige and gray, steel and glass. It looks like a prison town, whatever that means. Everywhere are signs stenciled with surveillance cameras. Everything here is being watched—every beige cube of a building; every dark window. And so this is a town where all the curtains seem to be closed—keeping something out, or protecting something within. To live here is to manage the cameras.

In the pedestrian alley that's devoid of pedestrians, a canopy of multicolored strips of crepe paper flutters over the concrete—remnants of some old celebration—their shadows dancing as if some maggot bacchanal. It's the maggots' time here now. The apartment complexes are prison-quaint, landscaped with bushes that appear to have once been neurotically trimmed but are now unkempt. They look uncomfortable, exhausted, pining for their old groomer, who either lost interest, or died, or got out of this place. The songbird life, though, is fabulous. In the courtyard of one apartment complex is a fountain—an iron sculpture of a giant toilet and a bathtub in which an aghast man sits, eyes wide, mouth open, arms resting on the sides of the tub. He's watching the toilet overflow, as his tub also overflows. Because he is iron, he will never not be aghast. In every apartment window, the curtains are drawn against this.

Besides its low-rise urban planning, Albertslund is known for being the site where one of the strongest tornadoes in Danish history touched down, ripping the tiles off roofs, destroying two schools, uprooting trees, and catapulting backyard trampolines into the air. Now bicyclists rocket along the town's web of paths. The paths pass beneath the streets, the tunnels inked with arrows and lightning bolts, so many concise manifestos, commands, and entreaties—FUCK PUTIN; PAINT YOUR EXIT; NO ENTRANCE TO THE MARE NEST. Above these tunnels, the trees are ratty feathers, and the weeds are the largest of their species. They have adapted a leatheriness. They are thick-skinned, and so they can take it. Fruits decay on their branches before dropping to the grass, arousing the ants from their hills. This smell of sweet rot is in the air. The one person I see walking here has the word *Lost* tattooed on the back of her neck.

I park the rental car at a lot on the outskirts of the prison, and six crows who were occupying the space fly away. One first lands on the hood and scratches it with its right talon before taking off. It makes an asterisk in the paint—a crude sketch of Herstedvester Prison itself. The prison is surrounded by a forest-green gate, maybe twenty feet high, big cameras affixed to its upper barrier and next to the door cut into it. To the left of the door, a red fire hydrant hides behind two pink boulders. Through the fence bars, one can see the decorative flourishes mounted at the prison entrance, a big gray cloud into which the prison's name is stenciled—HERSTEDVESTER FAENGSEL—over which arc gray concrete molds of birds. And all along the side of the prison (the side visible through the fence, at least), as if ascending diagonally from the beige facade, is another long line of concrete birds, wings up and open, frozen in midflight, petrified just as they began taking to the air. I think they may be ducks. The prison is covered in them.

≈

The crows circle above the prison. Again, I tell myself I am here to interview Madsen against my better judgment. Underscoring the screams of the crows, I swear I can hear Scott Waters's voice, once again intoning, "I can't help myself. I can't stop."

I can't stop. Though my vision blurs, and I can't feel my fingers or my toes, I somehow exit the car. My shoes crunch the gravel. I hold on to the car door to stay on my feet. I have left the lights on, and the car is letting me know, ringing its warning bell.

Herstedvester Prison opened its doors in 1935 for the detention of those deemed psychopathic and too dangerous to be held in the "open-style" prisons of Greenland. Outside of the confounding duck sculptures, the facade appears to have seen little upkeep since, rife with cracks, missing chunks in the shapes of cartoon explosions, great clouds of hardened bird excrement. Besides Madsen, these walls hold back the mass shooter responsible for Greenland's Narsaq massacre—the worst mass shooting in Greenland's history; a notorious Danish serial killer whose mother was his first known victim and who is infamous for filing lawsuits against journalists (and penning, from the prison, a bestselling book); bank robbers, hostage takers, and kidnappers.

In 2020, Madsen's access to the prison workshop was restricted after he fashioned therein a wooden "gun" that he used to threaten a therapist and warden in an attempted jailbreak. When he was recaptured five minutes later, five hundred meters from the prison, he lifted up his shirt and exposed an explosives belt. If the guards did not let him go, he said, he would detonate it. The belt, too, turned out to be fake, fabricated in the prison's carpentry workshop. Under Danish law, his sentence, of course, can't get any longer, but he isn't allowed to play with tools so much anymore, and his access to the prison library has been restricted as well.

I'm gripping the rental car door so hard it makes a red impression across my palm. I may be trying to delay my entrance into the prison. The car's lights are still on. The bell is still ringing. I don't want the bell to stop ringing. I don't want to confront the quiet, hear the blood in my ears or the sounds of my body. The crows circle. The prison looms. I can see how this is going to go. I have researched Herstedvester Prison and know what its visitation rooms look like. I want to encounter the fewest surprises possible. I know that the inmates sometimes get their haircuts in these rooms, so there may be tufts of hair on the floor. I know that there are plastic ferns on some of the tables; that the chairs are blue and purple; that a framed poster of the Eiffel Tower hangs crooked on a wall.

I can see it. I can see him—Madsen—in there, sitting at a low wooden table in one of the white-walled visitation rooms, his hands cuffed and in his lap, his baggy orange sleeves creased at the elbow.

I can see myself in there, too, as if from below: My hands shake as I pull out the blue chair across from Madsen and sit. He looks up at me from lowered brows, waiting for me to go first. My mouth is as dry as it's ever been. I ask him to tell me about the first time he remembered wanting to flee from the earth's surface in either a rocket or a submarine. My vision swims. I don't want to hear his voice—that flinty, reedy voice—echo from the concrete. Madsen opens his mouth, takes a breath. And Kim Wall, in her article about her visit to Brandenburg Gate, says, "I'm already soaked through from the autumn rain falling but two Frenchmen hold an umbrella over my head."

I nod, and my heart won't calm down, but I hear another question rush from me. "Some DIY submersible builders have described the sinking as becoming an addiction. Would you characterize it in that way?" Madsen wrinkles his nose, as if the question stinks, and it does. He curls his upper lip. And Kim Wall says, "This is not a day one should be building walls."

I ask Madsen, "Do you recognize any overlaps between the obsession with diving to depth and a penchant for committing violence?"

His handcuffs rattle in his lap, and his shoulders jump, and he thinks for minute. And Kim Wall says, "Happily chanting... *Wir sind ein Volk, we are one people*—the comrades around me break down the provisional fence, and their running feet are accompanied by laughter."

Madsen shrugs, mock-sheepishly I feel, and his cuffs rattle again. I ask if he still has violence in him. He leans back, smiles a little. I can hear the room itself beating, and I wonder if it's the heating oil, ticking into the radiator. The guard at the door paces, and I wonder if it's his footsteps. I wonder if it's coming from me, or from Madsen, or from an explosives belt hidden beneath his orange shirt. He leans forward. He opens his mouth. And Kim Wall says, "It all happens in the blink of an eye."

"Do you have regrets?" I ask Madsen, and he lifts his chin and I can see the stubble there, the rash there, and his teeth look too wet. And Kim Wall says, "I run with them ... The air is full of mist from the rain and the excitement."

And I ask him again, "Do you have regrets?" And Kim Wall says, "laughter," says, "rain," says, "excitement," says, "I run."

Madsen's hair seems to flatten, though he hasn't smoothed it. I think I may be having a heart attack. I decide to rephrase. "Are you sorry for what you've done?" He just opens his mouth like a slot. And, in so many text messages and voice mails and letters to her family and to her friends, Kim Wall says, "I love you." "I love you, I love you, I love you," she says.

≈

I can see all of this happening as I breathe in the parking lot, holding on to the car door, the lights still on, the alert bell still dinging. I let go of the car door. Finally having made it to this parking lot, I don't want it to happen. I am not built for this. I am out of my depth. I decide not to go in, to get out of here. I decide I can stop.

Between my shoes, the grass cracks through the asphalt, and I focus on the green of it, animated by something I can't see—the wind or a worm. I reach into my pocket for the car key, and it feels so good in my hand. It feels good to hold on to something. I look beyond the fencing, up that skinny road leading to and away from the prison. I turn my back on the place. Though it's not raining, a little geyser of mist appears in the air over the road. Quickly, it evaporates, but before it's fully dispersed, something that looks like a flaccid gray water balloon descends through it and plops onto the asphalt. In the foliage beyond the parking lot, the crickets start chirping, and their noise sends a chill up my spine. My hands tingle, and I turn back toward the prison, and I don't see a single person. I take my hand out of my pocket, let the key go. The crickets chirp, and I walk up the road toward the fallen balloon—the spot where that curtain of mist just was.

The balloon is not a balloon, but the carcass of a pacu, that invasive and carnivorous fish that plagues the Øresund Strait with its underbite, its humanoid teeth. It must be a juvenile—it's no bigger than my hand. I turn toward the prison. I want to know if anyone else has seen this, and can explain it to me. I turn back to the pacu. I squat to it. The air temperature drops, and the wind cools, and the leaves sound like static. I'm nervous, but not as nervous as I have been. Though the fish has rained down from the sky, there's nothing extraordinary about its body. This is not some biblical plague. This is not a miracle. It looks a little pathetic, really. There will be no annual festival inspired by its falling. The prison—formidable but crumbling—looms. Something electric turns on near the building and whirs. Somewhere, someone is rescuing a body from a shipwreck in the deep sea. Somewhere, someone is building an underwater house.

The crows reappear, circling. They're waiting for me to leave so they can eat. But I have to look first. I pick up a twig from the roadside grass and use it to open the pacu's mouth. Its lips make a slippery sound and open easily. And there they are, those big, square, humanoid teeth—the mouth that has begotten so many myths, so many nightmares. It's funny, in spite of everything I've read, it's not so horrible. Sure, it's arresting, and maybe even a little scary, but it also looks like us. Still squatting, I turn to the sky. It is gray. It is the color of the sea.

# EPILOGUE

At four hundred feet, the water is still clear. The surface light pinkens the reef, some dim wound-like throb hovering over the shadows of flying rays, schools of fish winking in and out as if stars, the languishing, condom-like pyrosomes bobbing at the whimsy of the water until stirred by a predator into swimming. When inspired to bioluminesce, the pyrosomes' openings dance as if a chorus of Munchian mouths, silently screaming beneath some disco ball. The bioluminescent crumbs—comprised of all sorts of creatures, crustaceans and dinoflagellates—once they appeared, never disappear. They are there like the cosmos, whirling through the viewport. Some flap their tiny wings. Orange, semitransparent squid push their way through the crumbs, jetting their plumes of purple ink into which blue thread-leg jellies take refuge.

*Idabel* is dropping into the sea, and I am sinking with it. From his captain's chair, over the rumble of the sub's motors, Karl Stanley tells me that he's known divers who become so bewitched at depth, partially because of the euphoria of nitrogen narcosis, that they unstrap themselves from their gear and dedicate their bodies to the sea. It can't get any better than this, they think. After feeling this, life on the surface could no longer be bearable, rife with the curdling wails of the sirens. Their remains are all over the place down here. The starfish overtake them.

Within the confines of the sub, at depth, Stanley's voice sounds hollow, flat. His words seem to stop short, unable to linger or echo. He speaks of the clarity he found when sinking to greater depths in

*Idabel*—both a metaphorical clarity and a literal one. Without the wind and waves agitating the water as they do closer to the surface, there's this thermocline at depth—a band of ocean wherein it seems as if one can see forever. "It's so *clean.*" And it is within this band that Stanley, seduced by this perceived purity, often decides to anchor *Idabel* on a fantastically shaped limestone shelf, and lie down, and breathe, and take inventory of his life before imagining that inventory also being wiped clean.

Stanley gushes about the rarely seen bluntnose sixgill sharks we're hoping to spot—the ones he pre-dropped that fermented bait for. Though they could damage the hull of thirteen-foot-long *Idabel*, the eighteen-to-twenty-foot-long shark's beauty, Stanley assures me, is worth the risk. They spend much of their time at five thousand feet. "But sometimes," he says, "they come up . . ."

We have long had names for these sharks—the Danish named them after combs, and the Greeks after the sharpness of their teeth. The Dutch named them after sundials, and the Spanish after cows. The French call them Sweet-mouth, and the Portuguese a Left-handed Gray. The Māori refer to them via the definitive exclamation "That's It." The Japanese—perhaps due to its size, or representation of the species—call it Shark Shark.

Perhaps it's wishful thinking because he loves them so much, but Stanley has been hatching a theory for over a decade that the sixgill shark is the most numerous creature of considerable size (which Stanley defines as two hundred pounds or more) with which we share the planet besides human-owned livestock. "I've bounced this theory off experts, and nobody can offer an alternative," he says. "It's reasonable to estimate there are more than a hundred million of these individuals . . . They've been left alone. They're too big for people to [kill] . . . It's not worth the hassle."

Sometimes, Stanley wants to see these sharks so badly that he'll haggle with local farmers to get livestock to use as lures—to, for example, "strap an eighty-pound pig to the side of the sub" to attract them. He has video of this, captured with a GoPro in a custom dry bag on the end of a carbon fiber pole. The images are lit in the red light one may expect in the back rooms of a nightclub dungeon, the dead little pig nodding forward and back as the shark bites into it, its gills waving open, deep as wounds. Stanley also shoots sick horses for bait. "[One] shark ripped the leg off a horse and swallowed it whole," he says. But before it swallowed, it swam away with the leg dangling from its jaws like a cigarette, the hoof knocking against its gills.

I picture *Idabel* as some undersea asteroid, whirling like one of those fairground swing rides, a bunch of carcasses attached to it, spinning out over the ocean floor. A horse whipping by, a pig...

When asked how he sees his obsession progressing, Stanley sighs, exasperated with the question, even as he's willing to provide answers. He conjures a reluctant sainthood, realizing that his extraordinariness also saddles him with the obligation to engage the plebes. He's had a sonar map made of the ocean features surrounding Roatán—erosion channels, undersea caves, boulder fields, knolls, abysses, pinnacles, waterfalls of sand. Stanley says that having this map is akin to having been a blind man who now can see an entire elephant for the first time. He plans to use this map to find "new communities of life."

He's recently designed a "sub-sled"—a floating sled made out of fiberglass—in order to tow *Idabel* out to more remote stretches of ocean at double the speed (taking it from three miles per hour to seven miles per hour) and stabilize the sub so he can get into and out of it in the turbulence at sea, an act which, in the pre-sled days, Stanley describes as "basically unsafe."

He thinks the future of submersible building may be leashed to 3-D printing. "Picture a bird's bone," he says, "and how it's a matrix of mostly space, but has an internal truss that's giving it incredible strength and light weight." He hopes that one day soon we may be able to 3-D print a submarine hull like this, out of titanium or aluminum, that may be able to go down to "almost full ocean depth" without imploding.

≈

I ask if there's been any downsides to his obsession. "There are definitely consequences," Stanley says. He's been a lifelong bachelor. "I've shaped my entire life around being close to deep water close to shore.

It's the number one reason why I'm on Roatán. It has definite shortcomings." He never has much money, isn't able to travel as much as he'd like; he doesn't have health insurance, and he wonders if he'll ever be able to retire. He says he meets a lot of people who, to his mind, haven't worked as hard, for as long, with as much passion, and who have a lot more to show for it. But it's the only way for him to live.

Roatán is part of the Mesoamerican Reef—the second largest in the world after Australia's Great Barrier Reef. At depth, Stanley says, we'll glimpse the bottom of this ancient reef upon which the entire island rests, bob amid constellations of iridescent shrimp, blennies, gobies and crabs, flamingo tongues, lettuce sea slugs, cowries and eagle rays, half-shark-half-rays called chimeras (which have no cartilage, and inspired *Idabel*'s logo), 120-foot-long jellyfish, dumbo octopuses that appear to swim with their ears, blue sea cucumbers elegant as ballroom dancers, gastropods that emit pink, yellow, and blue light, fish striped like tigers, and algae that whirl like the galaxies we've named after our gods. We might see columns of mirrorlike reflective fish that hang in the water heads-down, glass sponges that have survived four hundred million years, fish that look like three-foot flowers, fish with legs that prefer walking to swimming . . . If agitated by the whipping of a manta ray, or pod of dolphins, or whale shark, or hammerhead, glowing things might appear as if tossed confetti, fireworks of life rising and falling in the water, nudibranchs colliding with the seahorses. An octopus might cartwheel before getting its bearings to resume its hunt. Before uncoupling, the barracudas, jacks, and spadefish may find themselves swimming together. A squid could hold fast to the staghorn coral, its body fluttering like a moth's wings. The ostracods—the so-called bioluminescent "string of pearls"—having briefly dimmed, may reignite their fairy lights (thousands of strands hanging vertically in the water), resume their mating dance. Larval species will touch the bottom of the reef, as if a regenerative safe base, before retreating to the abyss. A frogfish might collide with a nurse shark, a moray eel with a pipehorse. Orange brain coral, coral resembling a field of purple roses, yellow branches of an endless quaking aspen, the pink bellies of upturned crab spiders, a wind-stirred savanna of swan feathers. We may see these things, Stanley says. Everything overdone and orphic, saturated and superlative. The worthwhile stuff. It's down here, he assures me.

~

At a thousand feet, the pink fades, the world goes dusky blue, a perpetual magic hour. Farther down, the ocean swallows natural light and goes pitch-black. In order for us to see these things, they must cross the beam of *Idabel*'s floodlight, where they appear as silvery snakes, siphonophorous balls of golden yarn, an army of petal-like wings, pulsing and spinning. We float with things that, after all of his diving, Stanley still can't name. "I love not knowing what you're looking at," he says. These things are so odd and unassimilable that I forget what they look like even as I'm looking at them.

"There are these prehistoric, filter-feeding relatives of starfish down here," Stanley says, "who are capable of walking around on the seabed. Most of the people who are aware of what they even are study fossils. They've been in the fossil record five hundred million years . . . To put it in perspective, dinosaurs were fifty or sixty million years ago. They have survived pretty much anything the planet and the universe can throw at them, and they're right here."

And right here, among them, at fifteen hundred feet, I'm shivering in my long sleeves. The condensation on the inner hull is frigid. It pools at my feet, dampens the toes of my socks. We meander amid a canyon of fallen coral boulders, where golden squat lobsters cling like spiders to sea-fan webs. Two-hundred-year-old orange roughy sniff the coral forests for the ghosts of their cousins who were whisked away to the diners of my childhood, gummed now by the ghosts of my long-dead grandparents who are still likely showing up for the Early Bird Specials of the Great Beyond. Polychaete worms writhe like chanterelle mushrooms belly-dancing. "Little yellow slimeballs," Stanley calls them.

We see colonies of glowing purple chandeliers and red umbrellas. As we approach, they break apart, before re-forming. Umbrellas become a red rain before becoming umbrellas again. A blue conger eel snakes from its lair in the coral, whips like a wind sock. It gums *Idabel*'s hull, tests our parameters, decides we're not edible, and, waving its tail, disappears into its hole. We see things we see straight through, and the ocean emerges strangely when glimpsed through the portals of their bodies, as if through some funhouse mirror. I'm being pulled into these bodies, my brain wormholing, and I have to grind my teeth, rub my temples to stay with myself. To keep from floating away into some fresh version of narcosis that will compel me to dedicate my body to the depth. I bite my tongue to the blood to prove my corporeality, remind myself that another kind of gravity exists and waits for us far above, where all that cruel, inspecting light is.

≈

Once, after being at depth for over seven hours on a shark dive, Stanley surfaced into a storm that roiled the ocean and knocked out the power on Roatán. The sub was nearly rolled over in the force of it, and as the shore lights were out, Stanley had to find the island in the darkness. He couldn't see land at all, and when the waves came up, they were so tall he couldn't see beyond them. He trusted his compass and made it home, but he says he still carries some of that terror with him. "Terrifying," he says, "fairly terrifying.

"There's been multiple windows cracking," on many of his dives, Stanley says, one of which resulted in "water spraying in the passenger's face." But this he laughs off. "We surfaced in a couple of minutes with maybe five gallons of water inside. I had the presence of mind to take his camera and wrap it in a towel. Nothing got damaged."

One may be tempted to consider, in this case, the schematics of damage. Whether a saved camera is enough to mute the trauma of nearly seeing one's ass handed to them at a depth of two thousand feet in an eminently crushable sub fashioned in an airplane hangar in rural Oklahoma. How would damage manifest in the aftermath of such a moment? I think of the ways—obvious and hidden—in which I am damaged. What damage cocktail might be responsible for my compulsion to test the qualities of an obvious damage (my irrational, dream-based fear of the ocean), by sinking into said ocean in this very vessel that forsook its own window, welcomed gallons of water into a previous passenger's face?

≈

At two thousand feet, the bioluminescent crumbs bunch together, and, for a moment, I can't see the ocean beyond them. Stanley says, "Here we go."

*Idabel* makes a shushing sound like white noise. It's taking something in or letting something go. I curl my toes in my wet socks. "What's happening?" I say. I breathe against the cold condensation on the steel, expecting to see my breath, but I don't. I wonder if I'm really here. I squeeze my forearm and it feels real enough—or like a memory of what I'm sure I used to feel like, up on the surface.

"Shark," Stanley says.

"Where?"

"There."

I can't see through the curtain of bioluminescence, but soon, it opens. I see the eye first, glowing like a cat's, then its face, its thick, sleek body gyrating. The water around it goes gauzy. The bioluminescence stills. The sixgill shark is silvery green, with a flattened head and rounded snout, and when it twists, the flesh at its gills fans open, then folds over on itself in cumulous bunches. It moves like the banner advertisement of my dream, falling from the Cessna to the fire. It bears the confidence of one who hasn't had to change a bit since the Triassic; a beast that realized something about survival—and perfection—before the rest of us. I reach for it, push my hand into the viewport's cave. I blink, and I regret blinking, because now it is only its tail, sweeping in a figure eight. And now, it is gone.

An orange octopus emerges from its hiding place. The ostracods twine like DNA. I want to say something to Stanley, test my voice, but I can't generate sound. I'm not sure why I'm about to cry. Stanley sighs, and it must be a loud sigh if I can hear it over *Idabel*'s drone. Out the viewport, green sperm-like creatures writhe on their journeys toward no earthly egg.

≈

Dramatically, but clearly rehearsed, Stanley counts down the last thousand feet of our fast ascent. "Four hundred feet," he says, and kills the motors, kills the lights, lets the air, fizzing into the ballast tanks, rocket us to the surface.

"Three hundred."

And in the darkest dark I've ever been immersed in, the bioluminescence agitates about the porthole, bounces from the Plexiglas like sparks from an oxyacetylene torch, some frenzied constellation thrashing, undoing itself, forsaking shape for chaos, some mosh-pit hyperspace.

"Two hundred."

*Idabel* is hot and humid again. The light from the surface begins to penetrate the water. The seagrass reappears on the coral walls and flashes as if with tea lights, the things living within it warning or clapping, or both, or neither. My heart accelerates, and my body remembers who I am, and though we're out of "certain death if something goes awry" territory, anxiety creeps back in.

"One hundred."

And I've got nothing to say.

We burst through the surface, foam pluming over the viewport, and there, muddied through the rinse of seawater, are the orange lights of town, human stuff. It's pushing 9:00 p.m. Soon, I'll hear the mopeds and diesels and honking horns along the beachfront road. I feel a great, implacable sadness, but I also want to laugh—cackle, actually. Stanley and I stay silent as we rumble toward his dock. I try to think about what happened. Sure, I experienced the obvious awe and wonder, but there was more to it. I also felt what I've only felt—in a slightly different way—at funerals. The unseen ethereal comprises so much more of our world than I had understood. I knew I would never be able to experience it all again, the way I knew I would never again hear the voices of those who had died. The experience was solemn and elegiac, and time cracked. Unstuck from, if not beyond, fear and anxiety, I felt aware of myself as a crumb on the continuum. And because I cannot bioluminesce, I felt invisible, and crushingly lonely, and already forgotten.

But mostly, I felt nothing. Utterly blank. Not just wiped clean but wiped away. I watched baby translucent orange squid strobe in the murk, plume their ink like mushrooms, and blissfully, as if already dead, I felt so quieted that I felt nothing.

≈

That night, at the motel, I'm alive and I'm having trouble sleeping, and trouble thinking. My head is full of the sea crumbs, and I want to see them again. I'm not sure what to do about this—what I'm going to do about this. I've sweated through the palm tree sheets, and I get out of bed, and do what I do at home to relax in the middle of the night. I open my laptop and read Emily Dickinson poems. In a desperate and futile attempt to contextualize the day, I search for ones about the sea. In poem 656, Dickinson begins: "I started Early—Took my Dog— / And visited the Sea— / The Mermaids in the Basement / Came out to look at me." The tide made as if He would eat the narrator up, "As wholly as a Dew," his "Silver Heel" upon her ankle. Her shoes "Would overflow with Pearl." At the end of the poem, "The Sea withdrew."

I begin to fade. I try again. I fall into bed. Soon, I am asleep. And I have my dream again. Still, my mother stretches out on the orange chaise longue. The propeller plane goes down, and the advertising banner chases it, and the beach grass shimmies, chocolate milk flying from my mother's thermos. And yes, the hotel again bursts into flames, as I'm pulled underwater, as I seek answers from the laconic eel as I drown. But somehow, something is different. Something elusive lurks beyond the eel, though it declares itself in no earthly way. It casts no shadow. It's not me who's withdrawing this time. Somehow—I can feel it—this drowning is the start of something, maybe my mother's fingers, once again corporeal, emerging to comb through my hair, as they did when I was a child, but maybe something else too. Something new. Something old. I don't know. And when I wake into the humidity of my motel room, clutching the damp sheets, I don't know exactly how I'm feeling. Maybe I'm waning, maybe I'm waxing. As if still underwater, I hear my heartbeat, louder than usual. Maybe I'm still afraid, but there's more to it. Some sort of hand has been extended. Hey: I'm still waking up. Outside the window, that's either the moon, or a wildfire. The palm fronds, or the woodpecker. Either way, the mermaids are out of the basement, and whether they're coming to look at me or not, my mother's father clings, overflowing with pearl, to their beautiful, dewy backs.

# Acknowledgments

Thank you:

To Kim Wall, whose work is forever, and rife with the sort of empathy and heart to which I will always aspire.

To the personal submersible enthusiasts who were willing to speak with me for this book, and who often endured my anxieties and my overuse of the word *obsession* with varying degrees of grace.

To all at the archives and libraries at the University of Copenhagen and Copenhagen's Saxo Institute, for your openness and access.

To Northern Michigan University, especially Rob Winn and David Wood, for their generous support of this project.

To Katie Ryder, editor extraordinaire at *Harper's Magazine,* for your careful eye, and for compelling me to face my fears and finally dive with Karl Stanley. The resulting article was published in the July 2023 issue of *Harper's,* some of which appears in this book, in slightly different form.

To Christopher Beha at *Harper's,* for green-lighting that article.

To Violet Lucca at *The Harper's Podcast,* for talking with me about amateur subs, the *Titan* disaster, obsession, wonder, language, and fear.

To my previous editors, Katie Henderson-Adams and Gina Iaquinta, whose voices constructively haunted me as I drafted this book, saving me so much time.

To my friends and colleagues who contributed support, feedback, and the occasional whisky to this endeavor: Elena Passarello, Jenny Boully, Aimee Nezhukumatathil, Rigoberto González, Doug Jones, Elizabyth Hiscox, Matt Bell, Christina Olson, Ben Drevlow, Sophfronia Scott, Benjamin Garcia, Donald Quist, Anna Clark, Leslie Contreras Schwartz, Dhonielle Clayton, S. Kirk Walsh, Jim Daniels, Karen Bender, Shonda Buchanan, Timston Johnston, Jen Howard, Jon Billman, Russ Prather, Carol Phillips, Alan Rose, Claire Rose, Jaspal Singh, Rafael Naranjo, Armando Santamaria, and . . .

To the amazing team at Pantheon Books: Andrea Monagle, Roland Ottewell, Carol Rutan, Annette Szlachta-McGinn, Marisa Nakasone, Tyler Comrie, Rose-Cronin Jackman, Bianca Ducasse, and Natalia Berry. You are rock stars, all.

To Lisa Lucas, for your generous support since the squid-book days.

To the continued support of my family: Mom, Noely, The Rub, Brian, Only Avery, Ella the Lizard.

To Maria Goldverg, for your enthusiasm, spot-on editorial prowess, and eagerness to make fun of awful music with me. This book is so much tighter and so much better for your advice.

To Rayhané Sanders, the best agent in the game, for . . . pretty much everything.

To my brilliant and hilarious editor Anna Kaufman, for seeing what, and how, this book needed to be; for your rigor, your open-heartedness, and your willingness to stay on the phone with me for four hours while a construction crew dismantled your kitchen ceiling. Every back-and-forth was a learning experience, and a joy.

And to you, Louisa. Yes: I promise to never lower myself into a submersible ever again.

# Bibliography

*ABS Rules for Steel Vessels for Vessels Certified for International Voyages*. United States Coast Guard, April 1, 2011.
Adan, Thoth. "Symbols Based on Circles." Thoth-adan.com, 2023.
"Advokatanpartsselskabet Per Bengtsson." Pbelaw.dk, 2021.
Aelian. *Varia Historia,* Chap. LXIV, penelope.uchicago.edu/aelian/varhist12.xhtml#chap64.
Ahonen, Marke. "Ancient Philosophers on Mental Illness." *History of Psychiatry* 30, issue 1 (March 2019): 3–18.
Albris, Laurine, and Martin Sejer Danielsen. "Tissø." Names in Denmark: Department of Nordic Studies and Linguistics, University of Copenhagen (names.ku.dk), October 10, 2014.
"All About Thalassophobia." PsychCentral.com, October 13, 2021.
Anderson, Christina, and Martin Selsoe Sorenson. "Danish Inventor Sentenced to Life in Prison for Reporter's Murder on Submarine." *New York Times,* April 25, 2018.
Androff, Susan. "Copenhagen's Latest Draw? An Abandoned Shipyard." *New York Times,* March 29, 2019.
"Aristotelian Concepts." Britannica.com (britannica.com/science/biology/Aristotelian-concepts).
"Aristotle (384–322 B.C.E.)," ucmp.berkeley.edu/history/aristotle.html.
"Aristotle and Marine Biodiversity." European Marine Board (marineboard.eu).
"Aristotle's Definition of Tragedy," tragedy.ucsc.edu.
Armstrong, Nancy, and Melissa Wagner. *Field Guide to Gestures: How to Identify and Interpret Virtually Every Gesture Known to Man*. Philadelphia: Quirk Books, 2003.
Arndt, Gary. "The History of the Submarine." Everything-everywhere.com, February 4, 2022.
"Arrian on Alexander the Great and Hephaestion." Ancientheroes.net, January 30, 2016.
Astrid Thomsen, Julie. "'Suicidal' Danish Submarine Owner Says Journalist Killed by Hatch Cover." Reuters, September 5, 2017.
Astrup, Peter. "Raket-Madsen er blevet forladt af sin kone." *B.T.,* February 11, 2018.
Ault, Alicia. "Renaissance Europe Was Horrified by Reports of a Sea Monster That Looked Like a Monk Wearing Fish Scales." *Smithsonian Magazine,* October 25, 2016.
Azhari, Timour. "Families of Migrant Boat Victims Find No Closure in Crisis-Ridden Lebanon." Reuters, September 1, 2022.

"Baby Submarine Prowls Streets and Searches Waters off Fidalgo Island." *Anacortes American* LXXV (August 20, 1964).

Bachrach, Arthur J. "History of the Diving Bell." *Historical Diving Times*, issue 21 (Spring 1998).

"Ban on Women on Submarines Ends." *New York Times*, February 24, 2010.

Baron, Kevin. "Navy Ending Ban on Women Aboard Subs." *Stars and Stripes*, February 24, 2010.

Bech Sillesen, Lene. "The Return." *Harper's Magazine*, September 17, 2020.

Beets, Jason. "Ready to Launch." *Salina Journal*, June 1, 2019.

Bennert, Vardha N., et al. "The Relation Between Black Hole Mass and Host Spheroid Stellar Mass out to z-2." Arxiv.org. 1102.1975, Cornell University, October 17, 2011.

Bennett, Paul. "The Pipe Dreamer." *National Geographic Adventure*, November 2002.

Berquist, Emma. "True Crime Is Rotting Our Brains." *Gawker*, October 12, 2021.

Bester, Cathleen. "Bluntnose Sixgill Shark: *Hexanchus griseus*." Florida Museum, University of Florida (floridamuseum.ufl.edu), July 19, 2021.

"Betina Hald Engmark Defends One of the Most Talked About Murder Suspects in Recent Times." DR P4 Radio Copenhagen, August 30, 2017.

"Black Holes, Quasars and Active Galaxies." Esahubble.org.

Blackwell, William. "Part Two: The Oral History of Saganaga and Northern Light Lake." *Cook County News Herald*, August 27, 2021.

Boas, Simon. "Jakob Buch-Jepsen bag facaden: Jeg lå søvnløs inden Peter Madsen-dommen." *B.T.*, March 30, 2019.

Bolin, Alice. "The Ethical Dilemma of Highbrow True Crime." *Vulture*, August 1, 2018.

Booth, Sebastian. "Does Distance or Remoteness Affect How Human Beings Use and Respond to Violence?" *E-International Relations* (e.ir-info), February 19, 2013.

Brennan, David. "U.S. Sailors Made 'Rape List' Aboard Submarine That Integrated Women but Tolerated 'Lewd and Sexist' Comments: Investigation." *Newsweek*, May 17, 2019.

"Briton Scoffs at Reputation of Submarine." *Indianapolis Times*, June 13, 1927.

Bronner, Simon J. *Crossing the Line: Violence, Play, and Drama in Naval Equator Traditions*. Amsterdam University Press, 2007.

———. "Sailor Men." National Sexuality Resource Center, San Francisco State University, (nsrc.sfsu.edu), June 18, 2007.

Buch, David. "Det kom frem på anden dag i retssagen mod Peter Madsen." Nyheder TV 2, March 21, 2018.

Burgess, Adam. "A Guide to Dante's 9 Circles of Hell." Thoughtco.com, November 1, 2019.

Burgess, Robert F. *Ships Beneath the Sea: A History of Subs and Submersibles*. New York: McGraw-Hill, 1975.

Calter, Paul. "Polygons, Tilings, and Sacred Geometry." Math.dartmouth.edu, 1998.

Campos Rivas, Orsy. "Lo que la lluvia regala a Yoro." *Revista Hablemos* (elsalvador.com/hablemos), July 17, 2004.

Carlson, Stephan C. "Golden Ratio." Britannica.com, 2023.

"Casualties: US Navy and Marine Corps Personnel Killed and Injured in Selected Accidents and Other Incidents Not Directly the Result of Enemy Action." Naval History and Heritage Command (history.navy.mil), December 7, 2022.

Chehayeb, Kareem. "Lebanese Submarine Finds 10 Bodies on Sunken Migrant Ship." Associated Press, August 26, 2022.

Chin, Richard. "What's Orange, Weighs a Ton and Sinks?" *Twin Cities Pioneer Press*, August 28, 2009.

Christou, William. "Spanish Submarine Pisces VI Arrives in Lebanon to 'Investigate' Tripoli Migrant Wreck." *The New Arab,* August 18, 2022.
Clante, Caroline, and Katrine M. Rasmussen. "Peter Madsen varetægtsfængslet for uagtsomt manddrab." *Ekstra Bladet,* August 12, 2017.
Clare, John D. "Arrian on the Siege of Tyre." Johndclare.net/AncientHistory/Alexander_Sources5.html.
Clerici, Caterina. "'We Don't Want Her to Be Remembered as the Victim': Kim Wall's Parents on Telling Her Story." *Guardian,* July 4, 2020.
Cleveland, Jonathan. "The Nature of Design: The Fibonacci Sequence and the Golden Ratio." Clevelanddesign.com, September 24, 2020.
"CLV/DSV Vina." JD-Contractor A/S (jdcon.dk), June 12, 2018.
"Commemorating 20th Anniversary of Ehime Maru." Nippon TV News 24 Japan, February 10, 2021.
"Community Submersibles Project." Communitysubmersiblesproject.com, 2022.
Copenhagen Suborbitals. Copenhagensuborbitals.com.
"Danish Swimmers Escape Waters Fearing Killer Fish." University of Copenhagen via phys.org, August 12, 2013.
Davis, Col. W. Jefferson. "The Ace of the Pacific." *Evening Star,* June 29, 1927.
"Deep Ocean Exploration | Karl Stanley | TEDxSanPedroSula." TEDx Talks, YouTube, December 2, 2021.
"Deep Sea Exploration in Homemade Submarines." Talks at Google, YouTube, April 16, 2018.
Demuth, Gary. "Kansas Man Launches Custom-Built Submarine." *Topeka Capital-Journal,* September 22, 2013.
Digges, Charles. "Russia Orders Halt to Submarine Towing in Wake of K-159 Disaster." Bellona.org, August 31, 2003.
"Diving Deep: Exploring the Depths with Homemade Submarines." Coolest Thing, YouTube, June 19, 2022.
Dodd, Vikram. "MI6 Worker Murdered, Stuffed in a Bag and Dumped in a Bath." *Guardian,* August 25, 2010.
Doniger, Wendy. "The Submarine Mare in the Mythology of Shiva." *On Hinduism.* Oxford: Oxford University Press, 2015.
Donnelly, Elaine. "Reasons Women Aren't on Subs." *Stars and Stripes,* August 28, 2009.
Dorn, Lori. "The Dedicated Woman Who Rehabs Old Submarines to Let Herself and Others Find Joy in the Submersible Life." Laughingsquid.com, August 27, 2019.
Doss, Erika. "Hopper's Cool: Modernism and Emotional Restraint." *American Art* (Smithsonian Institution) 29, no. 3 (Fall 2015).
Drabble, Margaret. "Submarine Dreams: Jules Verne's Twenty Thousand Leagues Under the Seas." *New Statesman,* June 24, 2021.
DuFord, Darrin. "Off the Deep End in Captain Karl's Homemade Yellow Submarine." Narratively.com, March 13, 2015.
Durant, Jay. "Meet the B.C. Man Who Builds His Own Fully-Functional Submarines." *Global News,* October 21, 2021.
Dyrssen, Elvira Lagerström. "Madsens fantasier: Om 'galna kvinnor' och dod." *Expressen,* March 23, 2018.
Eichinger Ferro-Luzzi, Gabriella. "Water in Modern Tamil Literature." *Revista degli studi orientali* 70 (1996): 353–66.
Ellwood, Mark. "The Unexpected Copenhagen Neighborhood Not to Be Missed." *WSJ Magazine,* October 5, 2017.

El-mochantaf, Christer. "Han var min allra basta van." *Expressen,* March 26, 2018.
Evans, Arthur B. "Hetzel and Verne: Collaboration and Conflict." *Science Fiction Studies* 28, part 1, no. 83 (March 2001).
Evans, David. "History." Bay Islands Diver (bayislandsdiver.com), August 6, 2020, https://web.archive.org/web/20161020205904/http:/bayislandsdiver.com/History.html.
Farah, M. J. "Perception and Awareness After Brain Damage." *Current Opinion in Neurobiology* 4, no. 2 (April 1994): 252–55.
"Finn Bachmann," finnbachmann.dk.
Fleckner, Mette. "Bag tremmer I vestre: Her er peter madsens liv I faengslet." *Ekstra Bladet,* August 25, 2017.
Gianotti, Carla. "A Note on the An Ityarthaparikatha." *Tibet Journal* 3, no. 2 (Summer 1988): 23–30.
Gilbert, Cecile. *International Folk Dance at a Glance.* Minneapolis: Burgess Publishing Company, 1974.
Gottipati, Sruthi. "My Friend Kim Wall's Disappearance in Denmark Shows: Female Journalists Face Danger Everywhere." *Guardian,* August 20, 2017.
"Gray Creek Claims to Be 'Home of the Gold Boulder.'" Waymarking.com, July 21, 2007.
Greve, Victoria. "Jag tanker pa Kim hela tiden." *Expressen,* August 16, 2017.
Haliburton, Rachel. "The Ethicist: Murder Mysteries and Moral Imagination." *Cape Breton Spectator,* March 29, 2017.
"Hall of Fame." Psubs.org.
Hammer, Joshua. "Vladimir Lenin's Return Journey to Russia Changed the World Forever." *Smithsonian Magazine,* March 2017.
Hancock, Paul. "BBC Lagos." Shipwreck Log (shipwrecklog.com), August 4, 2018.
"Hand Anatomy." Your Practice Online (ypo.education), 2023.
"Hangaren." Hangaren.dk.
Harkins, Gina. "Sailors Created 'Rape List' Aboard Navy's 2nd Sub to Integrate Women." Military.com, May 17, 2019.
Harper, Tyler. "Dangerous Oasis: The Fatal History of a Popular Kootenay Lake Beach." *Nelson Star,* October 14, 2020.
Health Plus. "Here's the Science Behind Why We Fall in Love." Her World (herworld.com), February 3, 2018.
Helgason, Gudmunder. "U-boat Myths and Stories." Uboat.net.
Hempsall, Vince. "Kootenay Man to Search for Gold with His Hand-Built Submarine." *Kootenay Mountain Culture Magazine,* no. 41 (May 2022).
Hendrikx, Sophia. "Fantastic Beasts and How to Make Them (According to 16th Century Instructions)." *Leiden Arts in Society Blog* (leidenartsinsocietyblog.nl), Universiteit Leiden, 2018.
"Hephaestion's Death and Funeral." Theworldofalexanderthegreat.wordpress.com, July 13, 2012.
Hirai, Masahiro, and Yasuhiro Kanakogi. "Communicative Hand-Waving Gestures Facilitate Object Learning in Preverbal Infants." *Developmental Science* 22, no. 4 (July 2019).
"History of Roatan." Coconuttreedivers.com.
Holland, Taylor Mallory. "Facts About Touch: How Human Contact Affects Your Health and Relationships." Dignityhealth.org, April 28, 2018.
Holowchak, Mark A. "Aristotle on Dreaming." *Ancient Philosophy* 16, issue 2 (Fall 1996): 405–23.
"Honduras—Into the Abyss in a Homemade Submarine." *Ramblings About the World* (ramblingsabouttheworld.com).

"How Do Hands Work?" National Library of Medicine: National Center for Biotechnology Information (ncbi.nlm.nih.gov), July 26, 2018.
Howie, Cary. *Claustrophilia: The Erotics of Enclosure in Medieval Literature*. New York: Palgrave Macmillan, 2007.
Impey, Chris. "3 Reasons Why Black Holes Are the Scariest Things in the Universe." *Astronomy*, November 12, 2020.
Irwin, John T. *Hart Crane's Poetry: "Appollinaire Lived in Paris, I Live in Cleveland, Ohio."* Baltimore: Johns Hopkins University Press, 2011.
Isachenkov, Vladimir. "Russia Bans Towing of Nuclear Subs After Sinking." *Irish Examiner*, September 1, 2003.
"Is It a Good Idea to Allow Women to Serve on U.S. Navy Submarines?" American Legion (legion.org), September 2, 2011.
"Jenny Hanivers, Mermaids, Devil Fish and Sea Monks." *Journal of the Bizarre* (bizarrejournal.com), May 23, 2012.
Jeong, May. "The Final, Terrible Voyage of the *Nautilus*." *Wired*, February 15, 2018.
Jespersen, Linette K. "Peter Madsen's Friend in Shock." *Ekstra Bladet*, August 24, 2017.
Jog, Chanda J., and Aparna Maybhate. "Measurement of Non-axisymmetry in Centres of Advanced Mergers of Galaxies." *Monthly Notices of the Royal Astronomical Society* 370, issue 2 (August 2006): 891–901.
Jonassen, Wendi, director. "How This Woman Started Diving in DIY Subs." *Wired*, video series *Obsessed*, season 1, episode 18, August 20, 2019.
Jonsson, Patrik. "A Submarine Sinks Myths About the Confederacy." *Christian Science Monitor*, July 25, 2001.
"Jules Verne, 'Twenty Thousand Leagues Under the Sea.'" Royal Museums Greenwich (rmg.co.uk), June 1, 2007.
Kalkomey. "The Early History of Submarines." Boaterexam.com, May 4, 2011.
———. "The History of Deep-Sea Exploration." Boaterexam.com, June 3, 2011.
"Kansas Historical Society: Fiscal Year 2013 Annual Report." Kshs.org, 2013.
"Kansas Man Working to Rebuild Old Submarine." KSHB 41, YouTube, May 27, 2016.
Kashef, Neosha S. "The Effects of Underwater Pressure on the Body." TED-Ed, YouTube, April 2, 2015.
Kelly, Bryan. "What the Politics of '20,000 Leagues Under the Sea' Mean." Inverse.com, September 24, 2015.
"Keynote Speaker: Commander Scott Waddle • Presented by SpeakInc • Failure Is Not Final." SpeakInc, YouTube, November 4, 2015.
"Kim Wall, Slain Journalist, Is Remembered in Her Own Words." *New York Times*, October 11, 2017.
"Kim Wall: What We Know About Danish Submarine Death." BBC News, April 25, 2018.
King, R. A. H. "Aristotle on Distinguishing *Phantasia* and Memory." In Fiona Macpherson and Fabian Dorsch, eds., *Perceptual Imagination and Perceptual Memory*. Oxford Academic (online edition), June 21, 2018.
Kosmetatou, Elizabeth. "The Aftermath: The Burial of Alexander the Great." Hellenic Electronic Center (greece.org), 1998, web.archive.org/web/20040827134332/http:/www.greece.org/alexandria/alexander/pages/aftermath.html.
———. "The Location of the Tomb: Facts and Speculation." Hellenic Electronic Center (greece.org), 1998, web.archive.org/web/20040531025749/http:/www.greece.org/alexandria/alexander/pages/location.html.

Kummel, B. "Nautiloidae-Nautilida." *Treatise on Invertebrate Paleontology.* Lawrence: Geological Society of America / University of Kansas Press, 1964.
"La Chambre Bleue." Center for Ocean Engineering, May 5, 2022.
Lahanas, Michael. "Alexander the Great and the Bathysphere." Hellenicaworld.com.
Lansberg, Paul. "Ictineo I, the First Ever Submarine." Marine Museums, Ships and Ports, lpsphoto.top/en/marine/barcelona-maritime-museo-iceneu-1, 2019.
Larsson, Lars. "Mesolithic Settlement on the Sea Floor in the Strait of Oresund." Lund University, conference paper, January 1983.
Larsson, Simon. "The Finds Inside the Submarine Can Trap Madsen." *Expressen,* April 3, 2018.
———. "The Women Tell About Peter Madsen's Life." *Expressen,* March 21, 2018.
"The Launch of UC3 Nautilus." Vimeo, May 5, 2008.
Lett, Phoebe. "Is Our True-Crime Obsession Doing More Harm Than Good?" *New York Times,* October 28, 2021.
Lim, Dion. "Stolen Homemade Submarine Now on Dry Land in Oakland." ABC 7 News (abc7news.com), April 30, 2018.
Lisagor, Kimberly. "Do-It-Yourself-Ahoogah." *Los Angeles Times,* April 6, 2004.
"Live Simply." CPHVillage.com, retrieved 2023.
"Lluvia de Peces (Rain of Fish)." Atlas Obscura (atlasobscura.com), June 28, 2010.
Lomholt, Anders. "Nye oplysninger fra Kim Walls kæreste beskriver de sidste timer op til den skæbnesvangre sejlads." Nyheder TV 2, January 25, 2018.
Love, William, and Nick Miroff. "Unclaimed Dead Get Burial at Sea." *East Bay Times,* August 17, 2016.
Luhn, Alec. "Russia's 'Slow-Motion Chernobyl' at Sea." BBC, September 1, 2020.
Macedo, Diane. "Women to Start Serving on Submarines, but Not Everyone's On Board." Fox News, December 23, 2015.
MacKinnon, Mark. "Sole Sub Survivor Under Guard in Hospital." *Globe and Mail,* September 3, 2003.
Martinez, Jolanie. "20 Years After Ehime Maru Tragedy, Former Navy Commander Apologizes in New Letter." *Hawaii News Now,* February 11, 2021.
Maybhate, Aparna, et. al. "The Formation of Spheroids in Early-Type Spirals: Clues from Their Globular Clusters." *Astrophysical Journal* 721, issue 1 (September 2010): 893–900.
Medeiros, João. "Grounded Submarine Photographed with Sonar." *Wired,* April 1, 2010.
Mette Lundofte, Anne. "The Kim Wall Murder Trial: The Case Against Peter Madsen." *New Yorker,* April 17, 2018.
———. "The Peter Madsen Guilty Verdict Leaves Lingering Questions and Pain." *New Yorker,* May 4, 2018.
Miller, Ron. "Everything You Know About Jules Verne Is Probably Wrong." *Gizmodo,* June 17, 2013.
"The Ministry of Defense Will Lift Submerged Nuclear Submarines from the Bottom of the Sea." Top War (en.topwar.ru), October 11, 2012.
Moscatello, Caitlin. "Disturbing Report: Female Naval Officers Allegedly Videotaped While Showering on Submarine." *Glamour,* December 5, 2014.
Moskowitz, Andrew K. "Dissociative Pathways to Homicide: Clinical and Forensic Implications." *Journal of Trauma and Dissociation* 5, issue 3 (2004): 5–32.
"Mother Published Book on Murdered Swedish Journalist Kim Wall." Thelocal.se, November 11, 2018.

"Murdered Journalist Kim Wall's Home Town Hosts Marathon in Her Memory." Getty Images (gettyimages.com).
Myers, Meghann. "Navy: Women Secretly Filmed in Shower Aboard Sub." *Navy Times,* December 3, 2014.
"The Myth of the Submarine by Ben Cockett." Toadgallery.com, 2022.
Nadeau, Barbie Latza. "Danish Scientist Charged with Beheading Journalist Wanted to Make a Snuff Film, Worried He Was a Psychopath." *Daily Beast,* March 28, 2018.
"Naval Myths and Traditions." Submarine Force Library and Museum Association (ussnautilus.org), September 15, 2017.
Nilsson, Lennart. "Allt du behöver veta om Kim Wall-fallet." *Expressen,* August 11, 2017.
"Note from Brazilian Congressman Herbert Levy." Wilson Center Digital Archive, December 13, 1982, digitalarchive.wilsoncenter.org/document/note-brazilian-congressman-herbert-levy.
O'Callaghan, Jonathan. "This Titan Submarine Is One of Several Futuristic Projects NASA Is Funding." IFLScience, July 7, 2015.
O'Flaherty, Wendy Doniger. "The Submarine Mare in the Mythology of Siva." *Journal of the Royal Asiatic Society of Great Britain and Ireland,* no. 1 (1971): 9–27.
Ohlsson, Per-Ola. "They Should Go to Space—in Home-Built Rocket." *Expressen,* March 13, 2010.
Oleson, Steven. "Titan Submarine: Exploring the Depth of Kraken Mare." NASA.gov, July 2, 2015.
Oltermann, Philip. "Woman's Body Found as Danish Police Search for Missing Journalist." *Guardian,* August 22, 2017.
Orange, Richard. "A Copenhagen Killing: The Story Behind the Submarine Murder." *Guardian,* January 7, 2018.
———. "'I Feel No Hate': Kim Wall's Mother in New Book." Thelocal.se, November 13, 2018.
Osborne, James. "The Gold Boulder." James Osborne Novels (jamesosbornenovels.com), September 23, 2013.
Østergaard, Jens Peder. "Dykkeren elsker hav-eventyr—og freden i den midtjyske hedeskov." *Viborg Folkeblad,* August 1, 2020.
Paiella, Gabriella. "Inventor Accused of Killing Journalist Reportedly Texted About Murdering Someone on His Submarine." *The Cut,* February 15, 2018.
Penner, Cindy. "Remembering My Brother and the Six Others Who Drowned in Kootenay Lake, 57 Years Later." *Creston Valley Advance,* January 6, 2021.
Piccard, Jacques, and Robert S. Dietz. *Seven Miles Down.* New York: G. P. Putnam's Sons, 1961.
Plant, David. "Kepler and the Music of the Spheres." Skyscript.co.uk, Summer 1995.
Pleasance, Chris. "Submarine That Police Expected to Find Body in Is Empty." *Daily Mail,* August 13, 2017.
Popkin, Jim. "Authorities in Awe of Drug-Runners' Jungle-Built, Kevlar-Coated Supersubs." *Wired,* March 29, 2011.
"Prompt Salvage of Submarine." JD-Contractor A/S (jdcon.dk), 2017.
Pronk, Hank. "Deepest Diving Homemade Submarine in the World Built by Hank Pronk." YouTube, January 27, 2020.
———. "Dr. Cliff Redus and R-300 Personal Submarine." YouTube, March 18, 2022.
———. "First Manned Dive in E3000 Plus Interior Tour | Homemade Submarine." YouTube, September 18, 2020.

———. "Kootenay Lake Treasure Hunting in Homemade Submarine." YouTube, January 15, 2022.
"Pronk Searches for Treasure in Submarine." *Columbia Valley Pioneer*, September 11, 2015.
PSUBS.org. (Personal Submersibles Organization: Amateur Submarine Builders and Underwater Explorers.)
"PSUBS-MAILIST," psubs.org/mlist/archive/0211/msg00048.html.
"PSUBS-MAILIST," psubs.org/mlist/archive/0708/msg00197.html.
"PSUBS-MAILIST," psubs.org/mlist/archive/0908/msg00387.html.
"PSUBS-MAILIST," psubs.org/pipermail/personal_submersibles/2013-October/000881.html.
"PSUBS-MAILIST," psubs.org/pipermail/personal_submersibles/2017-August/013015.html.
"PSUBS-MAILIST," psubs.org/pipermail/personal_submersibles/2017-August/thread.html#13080.
"PSUBS-MAILIST," psubs.org/pipermail/personal_submersibles/2017-September/013338.html; psubs.org/mlist/archive/0708/msg00231.html.
"Psychological Issues and Diving." DAN: Divers Alert Network (dan.org), retrieved 2023.
Ravella, Peter, and Tyler Buckingham. "Dive, Dive, Dive." *American Shoreline Podcast*, coastalnewstoday.com, March 25, 2021.
Rayner, Gordon. "Was MI6 Spy-in-a-Bag Gareth Williams Killed by 'Secret Service Dark Arts'?" *Telegraph*, March 30, 2012.
Richter, Matthew. "The Quest for Depth." *The Stranger*, December 19, 2009.
Rogoway, Tyler. "The U.S. Navy Has a Critically Important Submarine Test Base Tucked Away in Alaska." *The War Zone*, August 15, 2018.
Roslyng Olesen, Thomas. "From Shipbuilding to Alternative Maritime Industry—The Closure of Danyard Frederikshavn in 1999." *Erhvervshistorisk Årbog* 62, no. 2 (2013).
———. "Transforming an Industry in Decline." CBS: Copenhagen Business School, conference paper, 2019.
Rufus, S. "Turned On by Tight Spaces." *Psychology Today*, May 2, 2012.
Rupprecht, Caroline. *Womb Fantasies: Subjective Architectures in Postmodern Literature, Cinema, and Art*. Evanston, IL: Northwestern University Press, 2012.
"Russian Defense Minister Halts Towing of Derelict Nuclear Subs." Associated Press via *Arizona Daily Sun*, August 31, 2003.
Rychla, Lucie. "Puffin Discovery Made in Vicinity of Washed-up Whale." *Copenhagen Post*, February 29, 2016.
Rydhagen, Maria. "Peter Madsen's Outbreak—After the Questions in Court." *Expressen*, March 26, 2018.
"Salvatore J. Indiviglia Collection of WWII and Vietnam Material, c. 1922–2008." Hofstra University Library Special Collections Department, December 3, 2008.
Sawatsky, Dr. David. "Is Diving Addictive?" *Diver* (divermag.com), January 17, 2012.
Scales, Helen. "Discovered in the Deep: The 'Forest of the Weird.'" *Guardian*, October 11, 2022.
"Scott Waddle: Former Commander of the USS Greeneville and Author of The Right Thing." Thespeakersgroup.com, 2023.
"Sealing the Hatch of the UC3 Nautilus Submarine." Vimeo, April 28, 2008.
Seelye, John. "Oceans of Emotion: The Narcissus Syndrome." *Virginia Quarterly Review* (Spring 1982).
"Shanee Stopnitzky," shaneestopnitzky.weebly.com.

Sharp, Richard. "Kansas Man Rebuilds a Deep-Sea Submarine." KSHB-TV (khsb.com), May 29, 2016.
Shepard, Leslie A., ed. "Sea Phantoms and Superstitions." In *Encyclopedia of Occultism and Parapsychology,* 2nd ed., vol. 3, *P–Z and Indexes.* Detroit: Gale Research Company, 1985.
Shields, Christopher. "Aristotle's Psychology." *Stanford Encyclopedia of Philosophy* (plato.stanford.edu), October 12, 2020.
Shute, Clarence. *The Psychology of Aristotle: An Analysis of the Living Being.* New York: Columbia University Press, 1941.
Sillesen, Lene Bech. "The Return." *Harper's Magazine,* September 17, 2020.
Sinicki, Adam. "The Thanatos Instinct." HealthGuidance.org, February 11, 2020.
Slavin, Erik. "Breaking into the Underwater Boys' Club: Sailor One of 12 Women to Be Submarine-Qualified." *Stars and Stripes,* July 21, 2009.
Slawson, Nicola, and Mark Brown. "Submarine in Missing Journalist Case Sunk on Purpose, Danish Police Say." *Guardian,* August 13, 2017.
Sondberg, Astrid. "Foelg retssagen mod Peter Madsen foerste dag." Nyheder TV 2, March 8, 2018.
———. "Læs livebloggen fra retssagen mod Peter Madsen." Nyheder TV 2, March 21, 2018.
———. "Nu begynder sagen mod Peter Madsen—det her kommer til at ske i dag." Nyheder TV 2, March 8, 2018.
Sondberg, Astrid, and Julie Bjorn Teglgard. "Peter Madsen erkender partering af Kim Wall—og kommer med ny forklaring." Nyheder TV 2, October 30, 2017.
Sondberg, Astrid, Julie Bjorn Teglgard, and Magnus Bjerg. "Interviewet med 'Raket'-Madsen, opgaven Kim Wall aldrig vendte hjem fra." Nyheder TV 2, September 6, 2017.
Steel, Carlos. "'The Soul Never Thinks Without a Phantasm': How Platonic Commentators Interpret a Controversial Aristotelian Thesis." In Benedikt Strobel, ed., *Die Kunst der philosophischen Exegese bei den spätantiken Platon- und Aristoteles-Kommentatoren.* Berlin: De Gruyter, 2019.
Steinbuch, Yaron. "Submarine Owner Comes Clean About Missing Journalist's Death." *New York Post,* August 21, 2017.
Stone, Maddie. "The Wacky, Risky World of DIY Submarines." *Gizmodo,* August 21, 2018.
Stoneman, Richard. *The Greek Alexander Romance.* New York: Penguin Classics, 1991 (reprint).
Stoner, Cameron. "Women in Submarines: 10 Years Later." America's Navy (navy.mil), June 25, 2021.
Stopnitzky, Shanee, and Alessandra Nölting. "Submarines for the Rest of Us: Personal Submersibles Coming to Maker Faire Bay Area." *Make* (makezine.com), May 6, 2019.
Stromberg, Maya. "Peter Madsens mystiska sms innan ubaten sjonk." *Expressen,* August 12, 2018.
Styblo, Robert W. K., director. "Copenhagen Suborbitals 2 (OV) Peter Madsen." YouTube, July 1, 2014.
"Sub Ends Lebanon Mission After Finding Sunken Migrant Boat." Associated Press, August 30, 2022.
"The Submarine Myth and the Ace of the Pacific." Theleansubmariner.com, April 28, 2022.
"Submarine Project Hull Strength." Engineer's Edge (engineersedge.com), July 2, 2015–February 14, 2018.
"Submarine UC3 Nautilus Sank Today. Female Reporter Is Missing." Abovetopsecret.com, August 11, 2017, abovetopsecret.com/forum/thread1181675/pg3.

"Submarines in Art." Submarine Force Library and Museum Association (ussnautilus.org), March 12, 2019.
"Submarines in Kansas. Yes, Kansas." KAKE ABC (kake.com), December 12, 2016, kake.com/story/33765900/submarines-in-kansas.
Swaminathan, London. "How Did Ancient Hindus Find Submarine Mountains?" *Tamil and Vedas* (tamilandvedas.com), September 30, 2018.
———. "What Is Vadamukagni-Submarine Fire?" *Tamil and Vedas* (tamilandvedas.com), September 21, 2012.
Tawil, Fadi. "Lebanese Families File Lawsuit Against Army for Boat Sinking." Associated Press, September 1, 2022.
"This Woman Forever Changed the Way the Navy Designs Its Ships." Morale Patch Armory, May 28, 2019.
"Thracian Dance." Yorku.ca.
"Timeline in Case of Journalist Kim Wall's Death on Submarine." Associated Press, April 24, 2018.
"Trident Nuclear Weapons System Q&A." Campaign for Nuclear Disarmament (cnduk.org), July 2017.
Truelsen, Mary-Louise. "Anklageren i Ubådssagen: 'Der var dage, hvor jeg var trist og ked af det, når jeg kom hjem.'" *Alt* (alt.dk), April 17, 2019.
"Two Years After the K-159 Tragedy: The Submarine Remains at the Bottom." Bellona.org, September 1, 2005.
Tyson, Donald, ed. *The Fourth Book of Occult Philosophy*. Woodbury, MN: Llewellyn, 2009.
"UC3 *Nautilus* Launch." SubSim Radio Room (subsim.com), April 28, 2008.
"Undersea and Air Navies Predicted." *Evening Star*, March 31, 1927.
Valleau, Natalie. "'Who Doesn't Want a Submarine?': B.C. Man Looks for Buyer Online for Homemade Deep Diver." CBC News (cbc.ca), October 20, 2020.
Venosa, Ali. "Breaking Point: How Much Water Pressure Can the Human Body Take?" *Medical Daily* (medicaldaily.com), August 13, 2015.
Vickhoff, Alexander. "Madsens plan: Skicka ut sig sjalv i rymden." *Expressen,* August 15, 2017.
"Vina." Marinetraffic.com, 2023.
Vincent, Isabel. "Inside Slain Journalist's Fateful Submarine Ride." *New York Post,* August 26, 2017.
"Vragdykker: Dybet giver mig ro." *B.T.,* November 12, 2018.
Walbank, Frank W. "Alexander the Great." Britannica.com, March 31, 2023.
Walker, Thayer. "Off the Deep End." *Outside,* June 6, 2008.
Wall, Ingrid, and Joachim Wall. *A Silenced Voice*. Seattle: Amazon Crossing, 2020.
Wall, Kim. "Ghost Stories." *Harper's Magazine,* December 27, 2016.
———. "The Weekly Package." *Harper's Magazine,* July 2017.
Wall, Kim, and Caterina Clerici. "Vodou Is Elusive and Endangered, but It Remains the Soul of Haitian People." *Guardian,* November 7, 2015.
Ward, Peter. *In Search of Nautilus*. New York: Simon & Schuster, 1988.
Weber, Toby. "Imaging Expert Research Submarine Escapes, Heart Development." Cullen College of Engineering, University of Houston (egr.uh.edu), September 24, 2013.
"'We Can Do It!'—Shipbuilding Women Invade the Charlestown Naval Yard." Boston National Historical Park (nps.gov), January 17, 2023.
Weigel, David. "Kasich Campaigns in Search of Voters Who Aren't Ready to Settle." *Washington Post,* February 29, 2016.
Weiss, Mike. "Bay Waters a Favorite Final Resting Place." SFGate, June 6, 2007.

Weiss, Piero, and Richard Taruskin. *Music in the Western World: A History in Documents.* New York: Schirmer, 1984.

"Well-Known Lawyer Threatened." Jellypages.com, April 9, 2019.

Werner, Ben. "Submarine Community Can't Meet Demand from Female Sailors." USNI News (news.usni.org), November 11, 2019.

"What Do a Submarine, a Rocket, and a Football Have in Common?" *Scientific American*, November 8, 2010.

"When Did Waving Become a Part of Human Interaction?" History.stackexchange.com, November 2, 2014.

"Why Women Are Not Allowed to Work on USN's Submarines." Defencetalk.com, February 16, 2007.

Wolfe, Natalie. "Danish Inventor Peter Madsen to Spend the Rest of His Life in Prison for Murder of Journalist Kim Wall." *Courier Mail*, April 26, 2018.

Wood, Rodger L., Louise Maclean, and Ian Pallister. "Psychological Factors Contributing to Perceptions of Pain Intensity After Acute Orthopaedic Injury." *Injury* 42, no. 11 (November 2011): 1214–18.

"World in Brief." *Washington Post*, September 1, 2003.

"World's Second-Largest Reef, Mesoamerican Reef, Now in Decline." *Japan Times*, February 15, 2020.

Yi, Dr. Richard. "The Effect of Psychological Distance on Willingness to Engage in Ideologically Based Violence." Gemstone Program, University of Maryland, College Park, 2016.

Young, Grace C. "Pisces VI Submarine Unveiling in Kansas." *Grace Under Pressure* (graceunderthesea.com), August 11, 2019.

———. "Submarines in Kansas?" *Grace Under Pressure* (graceunderthesea.com), November 17, 2016.

Zezima, Katie, Robert Costa, and Philip Rucker. "GOP Candidates on a Frantic Sprint to Slow Trump Before Fateful Super Tuesday." *Washington Post*, February 29, 2016.

ABOUT THE AUTHOR

MATTHEW GAVIN FRANK is the author of the nonfiction books *Flight of the Diamond Smugglers, The Mad Feast, Preparing the Ghost, Pot Farm,* and *Barolo,* as well as the poetry books *The Morrow Plots, Warranty in Zulu,* and *Sagittarius Agitprop.* His work appears widely in journals and magazines, including *The Kenyon Review, Harper's Magazine, The Paris Review, Guernica, The New Republic, The Iowa Review, Salon, Conjunctions, The Believer, Freeman's,* and the *Best American Travel Writing* and *Best American Food Writing* anthologies.

A NOTE ABOUT THE TYPE

This book was set in Arno. Named after the Florentine river that runs through the heart of the Italian Renaissance, Arno draws on the warmth and readability of early humanist typefaces of the fifteenth and sixteenth centuries. While inspired by the past, Arno is distinctly contemporary in both appearance and function. Designed by Adobe principal designer Robert Slimbach, Arno is a meticulously crafted face in the tradition of early Venetian and Aldine book typefaces. Embodying themes Slimbach has explored in typefaces such as Minion and Brioso, Arno represents a distillation of his design ideals and a refinement of his craft.

Typeset by Scribe,
Philadelphia, Pennsylvania

Designed by Marisa Nakasone